Lecture Notes in Physics

Volume 972

The Lecture Notes in Physics

The series Lecture Notes in Physics (LNP), founded in 1969, reports new developments in physics research and teaching - quickly and informally, but with a high quality and the explicit aim to summarize and communicate current knowledge in an accessible way. Books published in this series are conceived as bridging material between advanced graduate textbooks and the forefront of research and to serve three purposes:

- to be a compact and modern up-to-date source of reference on a well-defined topic.
- to serve as an accessible introduction to the field to postgraduate students and nonspecialist researchers from related areas.
- to be a source of advanced teaching material for specialized seminars, courses and schools.

Both monographs and multi-author volumes will be considered for publication. Edited volumes should, however, consist of a very limited number of contributions only. Proceedings will not be considered for LNP.

Volumes published in LNP are disseminated both in print and in electronic formats, the electronic archive being available at springerlink.com. The series content is indexed, abstracted and referenced by many abstracting and information services, bibliographic networks, subscription agencies, library networks, and consortia.

Proposals should be sent to a member of the Editorial Board, or directly to the managing editor at Springer:

Dr Lisa Scalone
Springer Nature
Physics Editorial Department
Tiergartenstrasse 17
69121 Heidelberg, Germany
lisa.scalone@springernature.com

More information about this series at http://www.springer.com/series/5304

Jeff Greensite

An Introduction to the Confinement Problem

Second Edition

 Springer

Jeff Greensite
Physics and Astronomy Department
San Francisco State University
San Francisco, CA, USA

ISSN 0075-8450 ISSN 1616-6361 (electronic)
Lecture Notes in Physics
ISBN 978-3-030-51562-1 ISBN 978-3-030-51563-8 (eBook)
https://doi.org/10.1007/978-3-030-51563-8

This Springer imprint is published by the registered company Springer Nature Switzerland AG.
The registered company address is: Gewerbestrasse 11, 6330 Cham, Switzerland

To the memory of my parents, Arthur and Carol Greensite.

Preface

This contribution to the Springer *Lecture Notes in Physics* series is intended to provide an overview of the confinement problem in non-abelian gauge theory, with a particular emphasis on the relevance of center symmetry and the lattice formulation and an introduction to the current research. The book is an expanded and updated version of lectures and review talks on the confinement problem that I have presented over the years, particularly at the Schladming Winter School in 2005 and the Cracow School of Theoretical Physics in 2009.

It is a pleasure to thank Maarten Golterman and Štefan Olejník for volunteering to read the first draft of this manuscript and for their many helpful suggestions.

In this second edition, I have tried to briefly summarize some promising developments in this field which have occurred in the decade subsequent to the appearance of the first edition; these include new results in (1) the caloron/dyon theory of confinement; (2) the effective Polyakov line potential from functional methods; (3) the Dyson–Schwinger approach to confinement in Coulomb gauge; (4) further tests of vacuum wave functional proposals; and (5) the conformal light front/AdS model of hadron structure. There are in addition two new chapters: one presenting the latest numerical results relevant to the center vortex theory of confinement and the other concerned with the varieties of confinement in theories with matter transforming (as in the standard model) in the fundamental representation of the gauge group.

San Francisco, CA, USA   Jeff Greensite
April 2020

Contents

About the Author

Jeff Greensite received his Ph.D. in 1980 from the University of California at Santa Cruz and held postdoctoral positions at UC Berkeley, the Universite Libre de Bruxelles, and the Niels Bohr Institute in Copenhagen, before joining the faculty at San Francisco State University in 1984. His research specialty is in theoretical high energy physics. He has published over two hundred research papers, mostly in the general area of quantum chromodynamics and lattice gauge theory, along with one textbook *(An Introduction to Quantum Theory, IOP Publishing).*

Introduction

The standard model of particle physics, put forward almost 50 years ago, has been so successful at explaining observations that there are few experimental hints, at least from traditional sources such as accelerator centers, that anything more is needed. This situation could still change if something beyond the standard mode emerges from the Large Hadron Collider at CERN; it is, after all, unlikely that the standard model is the end of the story. The Higgs sector seems a little unnatural, among other things. There is also a strong CP problem, there are many free parameters, there is no explanation of dark matter (let alone dark energy) or three fermion generations, and the theory does not accommodate gravity. A minor adjustment is also required to allow for neutrino masses. On the other hand, given this adjustment, there is little reason to doubt the validity of the standard model as a description of all the known elementary particles, and their interactions, down to a distance scale of at least 10^{-15} cm, and there is good reason to admire a theory which can describe such a wide range of phenomena so economically.

Because of the great importance of the standard model, and the central role it plays in our understanding of particle physics, it is unfortunate that, in one very important respect, we don't really understand how it works. The problem lies in the sector dealing with the interactions of quarks and gluons, the sector known as Quantum Chromodynamics or QCD. We simply do not know for sure why quarks and gluons, which are the fundamental fields of the theory, don't show up in the actual spectrum of the theory, as asymptotic particle states. There is wide agreement about what must be happening in high energy particle collisions: the formation of color electric flux tubes among quarks and antiquarks, and the eventual fragmentation of those flux tubes into mesons and baryons, rather than free quarks and gluons. But there is no general agreement about *why* this is happening, and that limitation exposes our general ignorance about the workings of non-abelian gauge theories in general, and QCD in particular, at large distance scales.

This book concerns the confinement problem, which deals with the behavior of gauge theories, and the force which is mediated by gauge fields, at large distances.

© Springer Nature Switzerland AG 2020

J. Greensite, *An Introduction to the Confinement Problem*, Lecture Notes in Physics 972, https://doi.org/10.1007/978-3-030-51563-8_1

Historically, the word "confinement" in the context of hadronic physics originally referred to the fact that quarks and gluons appear to be trapped inside mesons and baryons, from which they cannot escape. There are other, and possibly deeper meanings that can be attached to the term, and these will be explored in due course. Although the confinement problem is far from solved, much is now known about the general features of the confining force, and there are a number of very well motivated (although not yet compelling) theories of confinement which are under active investigation. In this volume I will try to give an overview of the main ideas in this field, their attractive features, and, as appropriate, their shortcomings.

Global Symmetry, Local Symmetry, and the Lattice

2

Abstract

Introducing the basics: gauge invariance and the lattice formulation, the Monte Carlo method, and the impossibility of breaking a local gauge symmetry spontaneously. The possible phases of a gauge theory. Wilson loops, and the concept of magnetic disorder.

Confinement in non-abelian gauge theory involves the idea that the vacuum state is disordered at large scales; our best evidence that this is true comes from Monte Carlo simulations of lattice gauge theories. So to begin with, I need to explain what is meant by

- a disordered state,
- a lattice gauge theory,
- a Monte Carlo simulation.

I also need to explain, since this is a book about the confinement problem, what is meant by the word "confinement." That, however, is a surprisingly subtle, and even controversial issue, which will be deferred to the next, and the next to last, chapters of this book.

2.1 Global Symmetry and the Ising Model

Let's begin with the concept of a disordered state, as it appears in the Ising model of ferromagnetism. We know that certain materials (e.g. iron) can be magnetized at low temperatures, but above a certain critical temperature, known as the *"Curie Temperature,"* the magnetic moment disappears, at least in the absence of an external field. The tendency to retain a magnetic moment at low temperature is due

© Springer Nature Switzerland AG 2020
J. Greensite, *An Introduction to the Confinement Problem*, Lecture Notes in Physics 972, https://doi.org/10.1007/978-3-030-51563-8_2

to an interaction between neighboring atoms in the solid, whose potential energy is lowered if the magnetic moments of the neighboring atoms are aligned. The Ising model is a simplified picture of this situation. We imagine that the solid is a cubic array of atoms, and that each atom can be in one of two physical states: "spin up" or "spin down," with the magnetic moment oriented in the direction of the spin. The Hamiltonian of the Ising model is

$$H = -J \sum_x \sum_{\mu=1}^{D} s(x)s(x + \hat{\mu}) , \tag{2.1}$$

where $s(x) = 1$ represents an atom at point x with spin up, $s(x) = -1$ represents spin down, and J is a positive constant.

At low temperatures, in any dimension $D > 1$, the system tends to be in an *"ordered state"*, meaning that most spins tend to point in the same direction. The magnetization (average $s(x)$) is non-zero. At high temperatures the system is in a *"disordered state"*, in which the average spin is zero, corresponding to a vanishing magnetization. According to the usual principles of statistical mechanics, the probability of any given spin configuration $\{s(x)\}$ at a particular temperature T is

$$\text{Prob}[\{s(x)\}] = \frac{1}{Z} \exp\left[-\frac{H}{kT}\right] \tag{2.2}$$

and

$$Z = \sum_{\{s(x)\}} \exp\left[\beta \sum_x \sum_{\mu=1}^{D} s(x)s(x + \hat{\mu})\right] \tag{2.3}$$

where $\beta \equiv J/kT$. Now observe that the Hamiltonian $H[\{s(x)\}]$, and the probability distribution $\text{Prob}[\{s(x)\}]$, are left unchanged by the transformation of each spin by

$$s(x) \rightarrow s'(x) = zs(x) \text{ where } z = \pm 1 . \tag{2.4}$$

Although the transformation with $z = +1$ does nothing to the spins, we include it because the two transformations $\{1, -1\}$ together form a group, known as Z_2. The operation (2.4) is called a *"global"* transformation , because every spin $s(x)$ at every location x is transformed in the same way, by the same factor z.

Now it is obvious that the probability distribution is invariant with respect to the global Z_2 transformations, i.e. $\text{Prob}[\{s(x)\}]=\text{Prob}[\{zs(x)\}]$, and therefore the average spin

$$\langle s \rangle = \sum_{\{s(x)\}} \frac{1}{N_{spins}} \left(\sum_{x'} s(x')\right) \text{Prob}[\{s(x)\}] \tag{2.5}$$

must equal zero. After all, to any given spin configuration $\{s(x)\}$ contributing to the sum, with some average spin s_{av}, there is another configuration with spins $\{-s(x)\}$, with average spin $-s_{av}$, which contributes with exactly the same probability. From this argument, it would appear that magnets are impossible. And in a sense that's true...*permanent* magnets, permanent at a finite temperature for infinite time, *are* impossible! Suppose spins are aligned, at some low temperature, with $s_{av} > 0$. There will be small thermal fluctuations, and the s_{av} will vary a little from configuration to configuration, but in general, at low temperatures, s_{av} will be positive for a very long time. A thermal fluctuation which would flip enough spins so that s_{av} becomes negative is very unlikely. Nevertheless, providing the number of spins N_{spins} is finite, and we wait long enough, at some point one of these vastly unlikely fluctuations will occur, and then s_{av} will be negative, again for a very, very long, yet finite time. Averaged over sufficiently long time scales, the mean magnetization is zero. But the time between such large fluctuations grows exponentially with the number of spins, and for real ferromagnets of macroscopic size the time between flipping the overall magnetic moment, just by thermal fluctuations, would certainly exceed the age of the universe.

So using (2.5) gives a result which is formally correct at low temperatures, yet wrong for all "practical" purposes. It is therefore useful to introduce an external magnetic field h

$$H_h = -J \sum_x \sum_{\mu=1}^{D} s(x)s(x+\hat{\mu}) - h \sum_x s(x)$$

$$Z_h = \sum_{\{s(x)\}} \exp[-H_h/kT] \tag{2.6}$$

so that $\langle s \rangle \neq 0$ at any temperature, and then consider what happens in the pair of limits where we take first the volume (number of spins) infinite, and then reduce h to zero

$$m = \lim_{h \to 0} \lim_{N_{spins} \to \infty} \frac{1}{Z_h} \sum_{\{s(x)\}} \left(\frac{1}{N_{spins}} \sum_{x'} s(x') \right)$$

$$\times \exp\left[\beta \sum_{x,\mu} s(x)s(x+\hat{\mu}) + \frac{h}{kT} \sum_x s(x) \right]. \tag{2.7}$$

In this pair of limits, done in the order shown, it is possible that $m = \langle s \rangle \neq 0$. When this is the case, we say that the Z_2 global symmetry is *"spontaneously broken."* The term means that despite the invariance of the Hamiltonian, an observable (such as average magnetization) which is not invariant under the symmetry can nevertheless come out with a non-zero expectation value. At high temperatures, even in the limits shown, the magnetization m vanishes. In that case we say that the Z_2 symmetry is unbroken, and the spin system is in the symmetric phase. In general, in the unbroken symmetry phase, the symmetry of the Hamiltonian is reflected in the expectation values: the expectation value vanishes for any quantity which is not invariant with respect to the symmetry group. In the broken phase, non-invariant observables can have non-zero expectation values in the appropriate infinite volume limit.

We also say that the low-temperature, broken phase is an *"ordered"* phase, while the high-temperature, symmetric phase is the *"disordered"* phase. In the case of the Ising model, the term "order" obviously refers to the fact that the spins are not randomly oriented, but point, on average, in one of the two possible directions. This is not true in the disordered phase, although if a given spin in a typical configuration is pointing up, its immediate neighbors are more likely to point up than down. This correlation, however, falls off exponentially with distance between spins. The quantitative measure is provided by the correlation function of the spins at two sites, loosely denoted "0" and "R", which are a distance R apart in units of the interatomic spacing. The correlation function is defined as

$$G(R) = \langle s(0)s(R) \rangle$$

$$= \frac{1}{Z} \sum_{\{s(x)\}} s(0)s(R) \exp\left[\beta \sum_{x,\mu} s(x)s(x+\hat{\mu})\right]. \qquad (2.8)$$

In the disordered phase $G(R) \sim \exp[-R/l]$, where length l is known as the *"correlation length"*. In the ordered phase, $G(R) \to m^2$ as $R \to \infty$. It turns out that in $D = 1$ dimension, a spin system is in the disordered phase at *any* finite temperature; there is no phase transition between an ordered and a disordered phase, i.e. no non-zero Curie temperature. The symmetry-breaking transition makes its appearance for any dimension greater than one.

The existence of (at least) two phases, broken and unbroken, ordered and disordered, is characteristic of the statistical mechanics of most many body Hamiltonians that are invariant under some global symmetry.

2.2 Gauge Invariance: The Unbreakable Symmetry

The spin system which was just described is an example of a lattice field theory. The number of dynamical degrees of freedom (spins, in the Ising model) is discrete; each field component interacts with only a few neighbors, and they are usually (but not always) arranged on a square, cubic, or hypercubic lattice, in $D = 2, 3, 4$ dimensions respectively, as illustrated in Fig. 2.1 for $D = 3$ dimensions. The points of the lattice are known as *sites*, the lines joining neighboring sites are *links*, and the little squares bounded by four adjacent links are known as *plaquettes*.

A *gauge transformation* is a position-dependent transformation of the degrees of freedom, in which the transformation can be chosen independently at each site. The trick is to associate the dynamical degrees of freedom with the links of the lattice,

Fig. 2.1 A lattice is a
discretization of spacetime.
Gauge fields are associated
with the links between lattice
sites

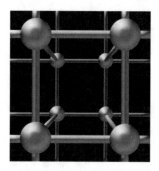

while the gauge transformation is specified at each site. A link can be uniquely
specified by a lattice site x and direction μ; the link then runs between sites x and
$x + \hat{\mu}$. The Hamiltonian for the gauge-invariant Ising model, also known as Z_2
lattice gauge theory, is

$$H = -J \sum_x \sum_{\mu=1}^{D-1} \sum_{\nu>\mu}^{D} s_\mu(x)s_\nu(x + \hat{\mu})s_\mu(x + \hat{\nu})s_\nu(x) .$$ (2.9)

The sum is over all plaquettes (Fig. 2.2), and the quantity which is summed is
the product of spin variables on links around each plaquette. This Hamiltonian is
invariant under the gauge transformation of each link variable $s_\mu(x)$ by

$$s_\mu(x) \rightarrow z(x)s_\mu(x)z(x + \hat{\mu}) ,$$ (2.10)

where the $z(x) = \pm 1$ can be chosen independently at each site; i.e. gauge
transformations are *local* transformations, and gauge symmetry is a local symmetry.
The local Z_2 gauge symmetry is vastly larger than the global Z_2 symmetry of the
Ising model, and this huge expansion of the symmetry can be shown to have the
following consequence [1]:

Elitzur's Theorem. *A local gauge symmetry cannot break spontaneously. The
expectation value of any gauge non-invariant local observable must vanish.*

This means there is no analog of magnetization; e.g. the expectation value
of a spin on a link $\langle s_\mu(x) \rangle$ is zero, even if we introduce an external field h
(which explicitly breaks the gauge symmetry) and then carefully take first the
infinite volume, and then the $h \rightarrow 0$ limits. We must look, instead, to gauge-
invariant observables, which are unaffected by gauge transformations. These can be

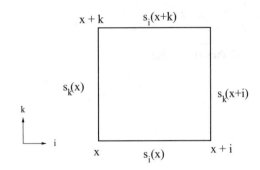

Fig. 2.2 A plaquette is the smallest loop on the lattice. The gauge field action is proportional to a product of link variables around each plaquette, summed over all plaquettes

constructed by taking the product of spins on links around a closed loop C, known as a *"Wilson loop"*[1]

$$W(C) = \left\langle \prod_{(x,\mu)\in C} s_\mu(x) \right\rangle . \tag{2.11}$$

The product of link variables around a plaquette, found in the gauge-invariant Hamiltonian (2.9), is just a particular example of a Wilson loop.

We can generalize the Z_2 construction to any symmetry group G, just by choosing link variables which are elements of G. Choosing G to be $SU(3)$, or $SU(2) \times U(1)$, we obtain the action of the gauge sector of the strong or electroweak interactions, in a discretized spacetime which allows for numerical simulations. A simple example is $G = U(1)$, and the corresponding gauge theory is known as *compact QED*. In this example the link variables are elements of the $U(1)$ group

$$U_\mu(x) = \exp[iaeA_\mu(x)] \quad , \quad A_\mu(x) \in \left[-\frac{\pi}{ae}, \frac{\pi}{ae}\right] , \tag{2.12}$$

where a is the lattice spacing, i.e. the distance (in meters, fermis, parsecs, or whatever) between neighboring sites on the lattice, and e is the unit of electric charge. The probability distribution of lattice gauge field configurations is

$$\text{Prob}[\{U_\mu(x))\}] = \frac{1}{Z}e^{-S[U]} , \tag{2.13}$$

where $S[U]$ is known as the "Euclidean action" (to be explained below), and is given by

$$S[U] = -\frac{\beta}{2} \sum_{x,\mu<\nu} U_\mu(x)U_\nu(x+\hat{\mu})U_\mu^*(x+\hat{\nu})U_\nu^*(x) + \text{c.c.} . \tag{2.14}$$

---

[1]The construction was first introduced by Wegner [2].

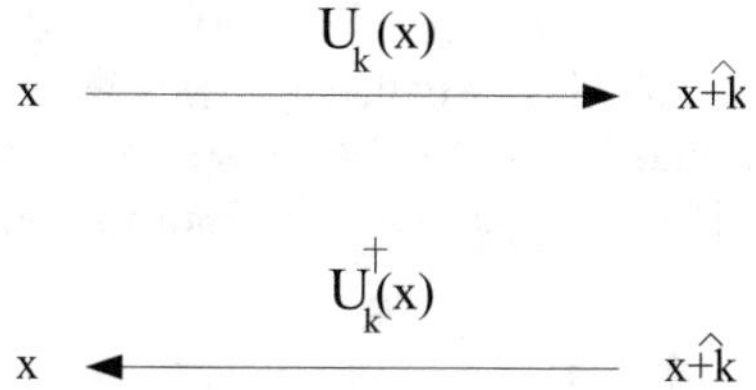

$$U_k(x)$$

x $\xrightarrow{\hspace{4cm}}$ x+$\hat{k}$

$$U_k^\dagger(x)$$

x $\xleftarrow{\hspace{4cm}}$ x+$\hat{k}$

Fig. 2.3 A gauge field variable $U_\mu(x)$ is represented graphically by a directed line joining points x and $x + \hat{\mu}$ in the positive μ-direction. The hermitian conjugate $U_\mu^\dagger(x)$ is represented by a line joining the same points, but in the opposite direction

It is customary to associate the link variable $U_\mu(x)$ with a line running from site x to site $x + \hat{\mu}$ in the positive μ-direction, and the complex conjugate $U_\mu^*(x)$ (or, for matrix-valued variables, the Hermitian conjugate $U_\mu^\dagger(x)$), with a line running from site $x + \hat{\mu}$ to x in the negative μ-direction, as shown in Fig. 2.3. In that case the product of link variables shown in (2.14), and its complex conjugate, can be visualized as Wilson loops running along the same plaquette in opposite directions. The Euclidean action $S[U]$ is invariant under local gauge transformations of the form

$$U_\mu(x) \to e^{i\theta(x)} U_\mu(x) e^{-i\theta(x+\hat{\mu})} \,, \tag{2.15}$$

where the set of transformations $g(x) = \exp[i\theta(x)]$ belongs to the compact $U(1)$ gauge group. To see why this theory is called compact QED, make the Taylor expansion

$$U_\mu(x) = 1 + iaeA_\mu(x) - \tfrac{1}{2}a^2e^2A_\mu^2(x) + \dots \tag{2.16}$$

Keeping terms in the action to lowest order in lattice spacing (and dropping an irrelevant constant)

$$S = \frac{\beta}{2} \sum_x a^4 e^2 \sum_{\mu < \nu} \left[\frac{A_\nu(x+\mu) - A_\nu(x)}{a} - \frac{A_\mu(x+\nu) - A_\mu(x)}{a} \right]^2 . \tag{2.17}$$

Identifying $\beta = 1/e^2$, we obtain in the limit $a \to 0$, known as the "*continuum limit,*"

$$S = \int d^4x \, \frac{1}{4} F_{\mu\nu} F_{\mu\nu} \,, \tag{2.18}$$

where $F_{\mu\nu}$ is the standard field strength of electromagnetism. However, in the action S the repeated indices are simply summed over, rather than contracted with a Minkowski metric. That is why S is termed a Euclidean action. It is obtained from the usual action of electromagnetism by rotating the time axis in the complex plane $t \rightarrow it$ (it is also necessary to rotate the time component of the vector potential, i.e. $A_0 \rightarrow -i A_0$). This is called a *"Wick rotation,"* and it essentially takes the metric from Minkowski to Euclidean space. In the continuum limit, the gauge transformation (2.15) is equivalent to the familiar form

$$A_\mu(x) \rightarrow A_\mu(x) - \frac{1}{e}\partial_\mu\theta(x) \; . \tag{2.19}$$

These constructions are readily generalized to non-abelian groups of interest, $SU(2)$ and $SU(3)$ in particular. For example, $SU(2)$ is the group of 2×2 unitary matrices of determinant one. We let the link variables be elements of this group

$$U_\mu(x) = e^{iagA_\mu(x)}$$
$$= b_0 \mathbb{1}_2 + i\mathbf{b} \cdot \boldsymbol{\sigma} \; , \quad b_0^2 + \mathbf{b} \cdot \mathbf{b} = 1 \; , \tag{2.20}$$

where $A_\mu(x)$ is a 2×2 Hermitian matrix

$$A_\mu(x) = A_\mu^a(x)\frac{\sigma_a}{2} \; , \tag{2.21}$$

and the $\{\sigma_a, \; a = 1, 2, 3\}$ are the three Pauli spin matrices. The Euclidean action

$$S = -\frac{\beta}{2} \sum_{x,\mu<\nu} \text{Tr}\left[U_\mu(x)U_\nu(x+\hat{\mu})U_\mu^\dagger(x+\hat{\nu})U_\nu^\dagger(x)\right] \tag{2.22}$$

is invariant under $SU(2)$ gauge transformations

$$U_\mu(x) \rightarrow G(x)U_\mu(x)G^\dagger(x+\mu) \; , \quad G \in SU(2) \; . \tag{2.23}$$

$SU(2)$ is a bit special, in the sense that the trace of any group element is real. The generalization to an $SU(N)$ gauge theory is simply to take the real part of the trace, i.e.

$$S = -\frac{\beta}{2N} \sum_{x,\mu<\nu} \left\{\text{Tr}\left[U_\mu(x)U_\nu(x+\hat{\mu})U_\nu^\dagger(x+\mu)U_\mu^\dagger(x)\right] + \text{c.c.}\right\} \; . \tag{2.24}$$

This is known as the *Wilson action*.

Once again, if we expand $U_\mu(x)$ in powers of $A_\mu(x)$, let $\beta = 2N/g^2$, and keep only terms which survive in the $a \rightarrow 0$ limit, we obtain

$$S = \int d^4x \; \frac{1}{2}\text{Tr}[F_{\mu\nu}F_{\mu\nu}] \; , \tag{2.25}$$

where this time $F_{\mu\nu}(x)$ is the non-abelian field strength

$$F_{\mu\nu} = \partial_\mu A_\nu - \partial_\nu A_\mu - ig[A_\mu, A_\nu] \; , \tag{2.26}$$

and the lattice gauge transformation (2.23) goes over to the continuum gauge transformation

$$A_\mu(x) \rightarrow G(x)A_\mu(x)G^\dagger(x) - \frac{i}{g}G(x)\partial_\mu G^\dagger(x) , \qquad (2.27)$$

with $G(x)$ an element of the gauge group.

2.3 The Monte Carlo Method

I have already mentioned that the lattice action S is a Euclidean action; spacetime indices are contracted with a Kronecker delta, rather than a Minkowski space metric. This Euclidean space formulation has a great advantage: the field theory can be regarded as a statistical (rather than quantum) system, and this lends itself to a powerful method of numerical simulation known as the lattice Monte Carlo method. Before describing that method, it is important to understand how a simulation in Euclidean space can yield results which are relevant to the corresponding quantum field theory in Minkowski space.

The connection between the Euclidean path integral, and the Minkowski space Hamiltonian operator H, is provided by the following relation: Suppose Q_t is some observable which depends on the fields at a fixed time t. Then

$$\langle Q_{t_2}^\dagger Q_{t_1} \rangle = \frac{1}{Z} \int DA \; Q_{t_2}^\dagger Q_{t_1} e^{-S}$$
$$= \langle \Psi_0 | Q^\dagger \; e^{-H(t_2-t_1)} \; Q | \Psi_0 \rangle , \qquad (2.28)$$

where the second line is an operator expression, with Ψ_0 the ground state of Hamiltonian H. With the help of this relation, it becomes possible to calculate properties of the Minkowski space theory, e.g. the low-lying spectrum, and the potential energy between static charges, from the lattice theory formulated in Euclidean space.

In a lattice Monte Carlo calculation, the idea is to replace the integral over all configurations $\{U_\mu(x)\}$, weighted by the probability distribution (2.13), by an average over a finite set of sample lattice configurations $\{U^{(n)}, \; n = 1, 2, \ldots, N_{conf}\}$, i.e.

$$\langle Q \rangle = \int DU \; Q[U] \frac{1}{Z} e^{-S[U]}$$
$$\approx \frac{1}{N_{conf}} \sum_{n=1}^{N_{conf}} Q\left[U^{(n)}\right] , \qquad (2.29)$$

where the sample configurations are generated stochastically, with probability weighting $\text{Prob}[\{U_\mu(x)\}] \sim \exp[-S[U]]$.

There are various ways to generate configurations with the desired probability weighting. The simplest and most general is known as the *"Metropolis algorithm,"* and the procedure is the following: Start with any convenient lattice configuration, e.g. all link variables set to the unit element. Sweep through the lattice link by link. At each link choose at random an element of the gauge group, denoted U', as a possible replacement for the existing link variable. Calculate the change ΔS in the action, if link variable $U_\mu(x)$ were replaced by U'. Then

1. If $\Delta S \leq 0$, set $U_\mu(x) = U'$.
2. Otherwise, for $\Delta S > 0$, use a standard random number generator to obtain a pseudo-random number x, in the interval $[0, 1]$ with uniform weighting. If $x < \exp[-\Delta S]$, set $U_\mu(x) = U'$. Otherwise, leave the link variable unchanged, and go on to the next link.

It can be shown that by iterating this procedure, one can generate as many lattice configurations as one likes with the desired probability weighting. Then any observable, such as a Wilson loop, can be averaged over the N_{conf} sample configurations to give an estimate of the expectation value. The statistical error inherent in this procedure goes down as $1/\sqrt{N_{conf}}$. While the Metropolis algorithm is easy to explain and easy to implement, it is often not the most efficient method in practice. For the more sophisticated algorithms which are used in modern, large-scale lattice Monte Carlo calculations, the reader should consult, e.g., the text by Gattringer and Lang [3].

We can use the expectation value of Wilson loops, computed in the Euclidean, statistical theory, to determine the interaction energy of two static, color-charged sources in the Minkowski space quantum theory. This interaction energy is known as the *static quark potential*, whether or not the sources are massive quarks. Consider, e.g., adding a massive scalar field to the gauge theory in D-dimensions, with action

$$S_{matter} = -\sum_{x,\mu}\left(\phi^\dagger(x)U_\mu(x)\phi(x+\hat{\mu}) + \text{c.c.}\right) + \sum_x \left(m^2 + 2D\right)\phi^\dagger(x)\phi(x) .$$
$$(2.30)$$

This action is gauge invariant, given that the scalar field transforms as $\phi(x) \rightarrow G(x)\phi(x)$. Let Q_t be an operator which creates a gauge invariant particle-antiparticle state at time t

$$Q_t = \phi^\dagger(0, t)U_i(0, t)U_i(\hat{i}, t)U_i(2\hat{i}, t)\dots U_i((R-1)\hat{i}, t)\phi(R\hat{i}, t) . \qquad (2.31)$$

For $m^2 \gg 1$ the one-link term in S_{matter} can be treated as a small perturbation. In this limit it is straightforward to integrate over ϕ in the functional integral, and we find, to leading order in $1/m^2$

$$\langle Q_T^\dagger Q_0 \rangle = \text{const.} \times \langle \text{Tr}[UUUU \dots UUU]_C \rangle , \qquad (2.32)$$

where $[UUU \ldots U]_C$ represents an ordered product of link variables (either U or $U^\dagger$), running counterclockwise around a rectangular $R \times T$ contour C, with the convention shown in Fig. 2.3. The trace of this quantity is a gauge-invariant Wilson loop, with expectation value denoted $W(R, T)$. The brackets $\langle \ldots \rangle$ in this case denote the expectation value of the observable in the gauge theory with no matter field, and the constant is $O(m^{-4(T+1)})$.[2] On the other hand, in the operator formalism we also have

$$\langle Q_T^\dagger Q_0 \rangle = \frac{\sum_{nm} \langle 0|Q^\dagger|n\rangle \langle n|e^{-HT}|m\rangle \langle m|Q|0\rangle}{\langle 0|e^{-HT}|0\rangle}$$

$$= \sum_n |c_n|^2 e^{-\Delta E_n T}$$

$$\sim e^{-\Delta E_{min} T} \quad \text{as } T \to \infty , \tag{2.33}$$

where the sum is over all energy eigenstates of the Hamiltonian containing two static charges at the given locations, and ΔE_n is the difference in energy of the n-th excited state and the vacuum. The minimal energy state containing two static charges, with energy difference ΔE_{min}, dominates the expectation value at large T, and this energy difference is also (apart from R-independent self-energy terms) the interaction energy between the static color-charged sources. Therefore, the relationship between the static quark potential $V(R)$ and the expectation value of the rectangular Wilson loop is

$$W(R, T) \sim e^{-V(R)T} \quad \text{(as } T \to \infty) . \tag{2.34}$$

2.4   Possible Phases of a Gauge Theory

Elitzur's theorem tells us that gauge symmetries don't break spontaneously, in the sense that any local, gauge non-invariant observable that could serve as an order parameter must have a vanishing expectation value. There is, nonetheless, a qualitative distinction between different phases of a gauge theory, based on the behavior of large Wilson loops. Let us again consider a large rectangular $R \times T$ loop, with $T \gg R$. There are three possibilities:

- **Massive phase.** The potential is of the Yukawa type

$$V(R) = -g^2 \frac{e^{-mR}}{R} + 2V_0 , \tag{2.35}$$

---

[2]This leading-order (in $1/m^2$) result is obtained by neglecting the one-link term in S_{matter} everywhere except along the timelike links from $t = 0$ to $t = T$, at $x = 0$ and $x = R$. On these links, expand $\exp[\phi^\dagger U_0 \phi + h.c.] \approx 1 + \phi^\dagger U_0 \phi + h.c..$ Integration over the scalar field then yields the result (2.32).

where V_0 is a self-energy term, and the Wilson loop falls off at $R > 1/m$ as $\exp(-2V_0T)$. For a generic, non self-intersecting planar loop C, of extension $\gg 1/m$,

$$W(C) \sim \exp[-V_0P(C)] \,, \tag{2.36}$$

where $P(C)$ is the perimeter of loop C.

- **Massless phases.** The potential has the form [4, 5]

$$V(R) = -\frac{g^2(R)}{R} + 2V_0 \,, \tag{2.37}$$

and the rectangular loop falls off as $\exp[-2V_0T + g^2(R)T/R]$. The effective coupling $g^2(R)$ can go to a constant (Coulomb phase), or, in gauge theories with massless matter fields, it may decrease logarithmically (free electric phase), or increase logarithmically (free magnetic phase), depending on the type of massless particles in the spectrum. For $R \gg V_0^{-1}$, the Wilson loop again has a perimeter-law falloff.

- **Magnetic disorder phase.** The potential is asymptotically linear

$$V(R) = \sigma R + 2V_0 \,. \tag{2.38}$$

Then the Wilson loop falls off asymptotically as $\exp[-\sigma RT - 2V_0T]$. More generally, for a planar non self-intersecting loop with minimal area $A(C)$ and perimeter $P(C)$,

$$W(C) \sim \exp[-\sigma A(C) - V_0P(C)] \,, \tag{2.39}$$

and there is an area-law type of falloff.

The term "magnetic disorder" is a terminology inspired by electromagnetism. Consider any lattice (or continuum) gauge theory, based on an abelian gauge group. Let C be a planar loop, whose minimal area $A(C)$ is subdivided into a set $\{C_i\}$ of n smaller loops, each of area A' as shown in Fig. 2.4. If the orientations of the loops are chosen so that neighboring contours run in opposite directions, it is easy to see that

$$U(C) = \prod_{i=1}^{n} U(C_i) \,. \tag{2.40}$$

Now suppose that if the subareas $\{A_i'\}$ are large enough, the expectation value of the product of loops is equal to the product of expectation values, i.e.

$$\left\langle \prod_{i=1}^{n} U(C_i) \right\rangle = \prod_{i=1}^{n} \langle U(C_i) \rangle \,. \tag{2.41}$$

Fig. 2.4 Subdivison of the minimal area of a large loop C into smaller areas bounded by loops $\{C_i\}$. In an abelian gauge theory, the Wilson loop around C is the product of Wilson loops around the $\{C_i\}$

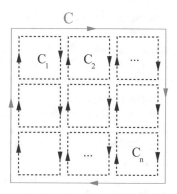

In that case,

$$W(C) = \langle U(C) \rangle$$
$$= \exp[-\sigma A(C)] , \tag{2.42}$$

where

$$\sigma = -\frac{\log[\langle U(C_i) \rangle]}{A'} . \tag{2.43}$$

If we consider varying the size and number of the subloops C_i, it is clear that, so long as the factorization property (2.41) holds, consistency requires that the string tension σ in (2.43) is independent of the subloop area A'.

The relationship (2.40), for the $U(1)$ gauge group, is simply a consequence of Stokes Law, since, in the continuum limit, for a planar loop in, e.g., the ij plane

$$U(C) = \exp\left[ie \oint_C dx^k A_k(x)\right]$$
$$= \exp\left[ie \int dS_C F_{ij}\right] , \tag{2.44}$$

where the second integration runs over the minimal area of the loop. Then (2.40) follows from the fact that

$$\int dS_C F_{ij} = \sum_n \int dS_{C_n} F_{ij} . \tag{2.45}$$

If i, j are spacelike directions, then in $3 + 1$ dimensions the integrals over S_{C_n} just give the magnetic flux passing through the area of the loop C_n.

It is Stoke's Law, in the abelian theories, that tells us that the product of smaller Wilson loops equals the single larger Wilson loop (2.40). The statement that the expectation value of the product of smaller loops is equal to the product of expectation values in (2.41) goes beyond Stoke's Law, and translates into the statement that if the minimal area A' is large enough, then the magnetic fluxes through the loops in the set $\{C_i\}$ are completely uncorrelated. This is the property which leads to an area law falloff for Wilson loops, and which motivates the term "magnetic disorder".

For non-abelian gauge groups, the Stokes law is not so simple (it exists, in fact, in several inequivalent versions, cf. [6–9]), but still boils down to the fact that a large planar Wilson loop can be expressed as the product of Wilson loops over smaller subareas.[3]

In $D = 2$ dimensions it is easy to demonstrate that the only phase that exists is the magnetically disordered phase, for any lattice action of the form (2.24), and for any gauge group. Take the Z_2 gauge theory for simplicity, with

$$S = -\beta \sum_p s(p)$$

$$s(p) = s_1(x)s_2(x + \hat{1})s_1(x + \hat{2})s_2(x) , \tag{2.46}$$

where the sum is over all plaquettes p defined on a rectangular lattice of N_p plaquettes, with free, rather than periodic boundary conditions.[4] We use the fact that, for $s(p)$ the product of links around a plaquette p,

$$\exp[\beta s(p)] = \cosh(\beta) + s(p)\sinh(\beta) , \tag{2.47}$$

and

$$\sum_{s_\mu=\pm 1} s_\mu = 0 , \quad \sum_{s_\mu=\pm 1} 1 = \sum_{s_\mu=\pm 1} s_\mu^2 = 2 . \tag{2.48}$$

Then for a loop C enclosing a region S of area $A(C)$

$$W(C) = \frac{\sum_{s_\mu(x)} \left(\prod_{(x',\mu') \in C} s_{\mu'}(x') \right) \prod_p (\cosh(\beta) + s(p)\sinh(\beta))}{\sum_{s_\mu(x)} \prod_p (\cosh(\beta) + s(p)\sinh(\beta))} . \tag{2.49}$$

---

[3]One approach, based on the non-abelian Stokes Law, derives an area law for a large Wilson loop from an assumed finite range behavior of field strength correlators, which means that field strengths are uncorrelated, i.e. disordered, at sufficiently large separations. This "field correlator" approach to magnetic disorder has been pursued by Simonov and co-workers [10].

[4]The Wilson loop calculation is a little easier in two dimensions with free boundary conditions. Periodic boundary conditions introduce a correction which is irrelevant for N_p large.

The only terms which survive in the numerator and denominator have either zero or an even number of link variables at each link, i.e.

$$\text{numerator} = \sum_{s_\mu(x)} \left(\prod_{(x',\mu')\in C} s_{\mu'}(x') \right) \prod_p (\cosh(\beta) + s(p) \sinh(\beta))$$

$$= \sum_{s_\mu(x)} \left(\prod_{(x',\mu')\in C} s_{\mu'}(x') \right) \prod_{p'\in S} s(p') \sinh(\beta) \prod_{p\notin S} \cosh(\beta)$$

$$= \sum_{s_\mu(x)} (\sinh(\beta))^{A(C)} (\cosh(\beta))^{N_p - A(C)}$$

$$= 2^{N_p} (\sinh(\beta))^{A(C)} (\cosh(\beta))^{N_p - A(C)} \, ,$$

$$\text{denominator}(= Z) = \sum_{s_\mu(x)} \prod_p (\cosh(\beta) + s(p) \sinh(\beta))$$

$$= \sum_{s_\mu(x)} (\cosh(\beta))^{N_p}$$

$$= 2^{N_p} (\cosh(\beta))^{N_p} \, , \tag{2.50}$$

so that

$$W(C) = (\tanh(\beta))^{A(C)}$$

$$= \exp[-\sigma A(C)] \quad \text{with} \quad \sigma = -\log[\tanh(\beta)] \, . \tag{2.51}$$

This demonstrates that Wilson loops have an area law falloff, and Z_2 lattice gauge theory is in the magnetically disordered phase, for any value of $\beta > 0$ in $D = 2$ dimensions. The calculation for $U(1)$ and $SU(N)$ gauge groups is similar, with $\cosh(\beta)$ and $\sinh(\beta)$ replaced by other special functions. For abelian gauge groups such as Z_2 and $U(1)$, it is also straightforward to verify the property (2.41); in the case of Z_2 gauge theory both sides of (2.41) equal $\tanh^{A(C)}(\beta)$. For non-abelian groups, an expression of the form (2.41) holds, in $D = 2$ dimensions, for the product of traces $\text{Tr}[U(C_i)]$ of loops.

The underlying reason for magnetic disorder in $D = 2$ dimensions is the absence of a Bianchi constraint relating different components of the field strength tensor. In general, the classical gauge field equations in $D = 4$ dimensions are of two sorts. First, there are the genuine equations of motion, which follow from variation of the action with respect to A_μ^a

$$(D^\mu)^{ab} F_{\mu\nu}^b = j_\nu^a \, , \tag{2.52}$$

where D_μ is the covariant derivative

$$D_\mu^{ab} = \delta^{ab}\partial_\mu - gf^{acb}A_\mu^c , \tag{2.53}$$

and j_μ^a is an external current source. In electromagnetism, where the color indices a, b are absent, and the covariant derivative is just an ordinary derivative, these equations of motion are Gauss's Law and Ampere's Law. But there are also a set of conditions on the field strength tensor which follow as identities from the fact that the field strength is expressed in terms of the 4-vector potential

$$F_{\mu\nu}^a = \partial_\mu A_\nu^a - \partial_\nu A_\mu^a + gf^{abc}A_\mu^b A_\nu^c . \tag{2.54}$$

These identities are the Bianchi constraint

$$(D^\mu)^{ab*}F_{\mu\nu}^b = 0 , \tag{2.55}$$

where $^*F_{\mu\nu}^b$ is the "dual" field-strength tensor

$$^*F_{\mu\nu}^a = \tfrac{1}{2}\varepsilon_{\mu\nu\alpha\beta}F^{a\alpha\beta} . \tag{2.56}$$

In electromagnetism, the Bianchi constraint in $D = 4$ dimensions consists of the source-free Maxwell's equations, i.e. Faraday's Law plus the condition $\nabla \cdot B = 0$. In $D = 3$ dimensions there is only the condition $\nabla \cdot B = 0$. In $D = 2$ dimensions the Bianchi constraint is absent.

Many years ago, Halpern [11] showed that it was possible to change integration variables in the functional integral from the gauge potentials $A_\mu(x)$ to the field strength variables $F_{\mu\nu}$, with the result that

$$Z = \int DA_\mu \, \exp\left[-\int d^D x \, \tfrac{1}{2}\mathrm{Tr}\left[F_{\mu\nu}^2\right]\right]$$
$$= \int DF_{\mu\nu} \, \delta[I(F)] \exp\left[-\int d^D x \, \tfrac{1}{2}\mathrm{Tr}\left[F_{\mu\nu}^2\right]\right] , \tag{2.57}$$

where $I(F) = 0$ is the Bianchi constraint, expressed entirely in terms of the (abelian or non-abelian) field strengths. In this formulation it is clear that the only thing which correlates the values of the field strength at neighboring points is the Bianchi constraint. In two dimensions this constraint is absent, and field strengths fluctuate independently from one point to another. A similar change of variables, from link to plaquette variables, can be carried out in lattice gauge theory [12], with the similar result that in two dimensions the plaquette variables fluctuate independently. Again using Z_2 lattice gauge theory as an example, we can use the gauge freedom to transform some of the link variables to $+1$:

$$s_1(x) = 1 \quad , \quad s_2(x_1 = 0, x_2) = 1 . \tag{2.58}$$

This is a choice of gauge, analogous to the Coulomb or Landau gauge-fixing conditions in electromagnetism. We may then change variables from the remaining link variables $s_2(x)$ to (an equal number of) plaquette variables $s(p) = s_2(x + \hat{1})s_2(x)$, so that

$$Z \to \prod_p \sum_{s(p)=\pm 1} (\cosh(\beta) + s(p)\sinh(\beta)) , \qquad (2.59)$$

and we see that the probability distribution factorizes into a product of independent distributions for each plaquette. Using the Stokes Law

$$\prod_{(x',\mu')\in C} s_{\mu'}(x') = \prod_{p\in S} s(p) \qquad (2.60)$$

we can quickly rederive the result for Wilson loops

$$\begin{aligned}
W(C) &= \frac{1}{Z} \prod_{p'\in S} \sum_{s(p')=\pm 1} s(p') \left(\cosh(\beta) + s(p')\sinh(\beta) \right) \\
&\quad \times \prod_{p\notin S} \sum_{s(p)=\pm 1} (\cosh(\beta) + s(p)\sinh(\beta)) \\
&= \frac{(2\sinh(\beta))^{A(C)}(2\cosh(\beta))^{N_p-A(C)}}{(2\cosh(\beta))^{N_p}} \\
&= (\tanh(\beta))^{A(C)} .
\end{aligned} \qquad (2.61)$$

It is clear in this case that the absence of a correlation among even the smallest loops (i.e. plaquette variables) is the crucial ingredient which leads to the area-law expression for Wilson loops of any size. The possibility of changing from link to plaquette variables, without any additional constraint on the plaquette variables, is unique to $D = 2$ dimensions.

In $D > 2$ dimensions, the area law falloff of Wilson loops holds at strong couplings, i.e. $\beta \ll 1$; the calculation, to leading order in β, gives the same result as in $D = 2$ dimensions. At weak couplings, $\beta \gg 1$, the area law certainly does not hold for small loop areas. This is a good thing; we would not like to end up predicting a linear potential between positive and negative electric charges in $3 + 1$ dimensions, and in fact one finds that $U(1)$ gauge theory is in the massless phase in $D = 4$ dimensions at weak couplings. But the area law falloff has been found, via Monte Carlo simulations, to hold for sufficiently large loop areas in non-abelian gauge theories.

In the case of the Ising model, the ordered and disordered phases were distinguished by the breaking of a global Z_2 symmetry. Elitzur's theorem assures us that the magnetically disordered phase, and the massive/massless phases, cannot be distinguished by the breaking of the local gauge symmetry. But is there some other global symmetry which would serve this purpose? We will turn to this issue in the next chapter.

References

1. S. Elitzur, Impossibility of spontaneously breaking local symmetries. Phys. Rev. D **12**, 3978 (1975)
2. F.J. Wegner, Duality in generalized Ising models and phase transitions without local order parameter. J. Math. Phys. **12**, 2259 (1971)
3. C. Gattringer, C.B. Lang, in *Quantum Chromodynamics on the Lattice*. Lecture Notes Physics, vol. 788 (Springer, Berlin, 2010)
4. N. Seiberg, Electric-magnetic duality in supersymmetric nonAbelian gauge theories. Nucl. Phys. B **435**, 129 (1995). arXiv:hep-th/9411149
5. K.A. Intriligator, N. Seiberg, in *Strings 95: Future Perspectives in String Theory*. Phases of N = 1 supersymmetric gauge theories and electric-magnetic triality. arXiv:hep-th/9506084
6. D. Diakonov, V.Y. Petrov, A formula for the Wilson loop. Phys. Lett. B **224**, 131 (1989)
7. R.L. Karp, F. Mansouri, J.S. Rno, Product integral formalism and non-Abelian Stokes theorem. J. Math. Phys. **40**, 6033 (1999). arXiv:hep-th/9910173
8. I. Arefeva, Nonabelian stokes formula. Theor. Math. Phys. **43**, 353 (1980). [Teor. Mat. Fiz. **43**, 111 (1980)]
9. P.M. Fishbane, S. Gasiorowicz, P. Kaus, Stokes' theorems for nonabelian fields. Phys. Rev. D **24**, 2324 (1981)
10. A. Di Giacomo, H.G. Dosch, V.I. Shevchenko, Y.A. Simonov, Field correlators in QCD: theory and applications. Phys. Rep. **372**, 319 (2002). arXiv:hep-ph/0007223
11. M.B. Halpern, Field strength and dual variable formulations of gauge theory. Phys. Rev. D **19**, 517 (1979)
12. G.G. Batrouni, M.B. Halpern, String, corner and plaquette formulation of finite lattice gauge theory. Phys. Rev. D **30**, 1782 (1984)

What Is Confinement?

3

Abstract

Confinement and symmetry. Linear Regge trajectories, the linear static quark potential, and the "spinning stick" model. String breaking, existence of "color confinement" in the Higgs phase, and the difficulty of associating confinement with an unbroken or spontaneously broken remnant gauge symmetry. Confinement as magnetic disorder; center symmetry as the global symmetry corresponding to the confined phase.

What is meant by the word "confinement"?

Most efforts aimed at proving (or at least understanding) confinement attempt to show that the potential energy between a static quark-antiquark pair (the *static quark potential*), grows linearly with the separation between the quarks. In other words, since a linear potential is associated with an area-law falloff for Wilson loops, the aim is to show that $SU(3)$ gauge theory in $D = 4$ dimensions is in the phase of magnetic disorder. But if, by the word "confinement," we mean that there is a static quark potential which rises indefinitely with quark separation, then by this definition QCD is not confining, because the static quark potential must eventually go flat due to a process, described below, known as "string breaking".

For this reason, since all of the known hadrons are singlets under the $SU(3)$ color group, many people prefer to define confinement as the absence of color charged asymptotic particle states. This defines confinement as "color confinement." Although this definition ties in very well with the historical understanding of confinement as the absence of free quarks, it has its own problems if taken seriously. The difficulty is that the asymptotic particle states of a gauge-Higgs theory, where the forces are all Yukawa-like and the dynamics something like the weak interactions, are *also* color singlet states. So if by "confinement" we mean color confinement, then Higgs theories are also confining.

© Springer Nature Switzerland AG 2020

21

J. Greensite, *An Introduction to the Confinement Problem*, Lecture Notes in Physics 972, https://doi.org/10.1007/978-3-030-51563-8_3

Before settling on a choice of terminology, at least for the purposes of this book, it is useful to understand *why* a linear static quark potential is believed to be relevant to QCD, why that potential must ultimately level off, and how we know that the spectrum of a Higgs theory consists only of color singlets.

3.1   Regge Trajectories, and the Spinning Stick Model

What is special about the dynamics of QCD, as opposed to, e.g., the electroweak theory? A very strong clue comes from the QCD spectrum, when the spin of mesons (and baryons) is plotted against their squared mass, as shown in Fig. 3.1. In such plots the mesons and baryons of given flavor quantum numbers seem to lie on nearly parallel straight lines, known as linear Regge trajectories. This is a very striking feature of the QCD spectrum, nothing similar is found in the electroweak theory, and the question is why it occurs.

Suppose that we picture a meson as a straight line of length $L = 2R$, with mass per unit length σ. The line rotates about a perpendicular axis through its midpoint, such that the endpoints of the line are moving at the speed of light. Then for the energy (mass) of the spinning stick we have

$$m = E = 2 \int_0^R \frac{\sigma \, dr}{\sqrt{1 - v^2(r)}}$$

$$= 2 \int_0^R \frac{\sigma \, dr}{\sqrt{1 - r^2/R^2}}$$

$$= \pi \sigma R , \tag{3.1}$$

Fig. 3.1 Regge trajectories for the low-lying mesons (figure from Bali [1])

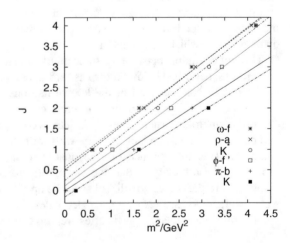

and for the angular momentum

$$J = 2 \int_0^R \frac{\sigma r v(r) dr}{\sqrt{1 - v^2(r)}}$$

$$= \frac{2}{R} \int_0^R \frac{\sigma r^2 dr}{\sqrt{1 - r^2/R^2}}$$

$$= \tfrac{1}{2} \pi \sigma R^2 . \tag{3.2}$$

Comparing the two expressions, we see that

$$J = \frac{1}{2\pi\sigma} m^2 = \alpha' m^2 . \tag{3.3}$$

The constant α' is known as the *Regge slope*. From the data, $\alpha' = 1/(2\pi\sigma) = 0.9 \, \text{GeV}^{-2}$, which gives a mass/unit length of the string, or *string tension*, of

$$\sigma \approx 0.18 \, \text{GeV}^2 \approx 0.9 \, \text{GeV/fm} . \tag{3.4}$$

The spinning stick model isn't perfect; in fact the various Regge trajectories shown in Fig. 3.1 do not pass through the origin, and have slightly different slopes. To make the model more realistic, one might want to relax the requirement of rigidity, and allow the "stick" to fluctuate in transverse directions. This line of thought leads to the formidable subject of string theory. But since QCD is the theory of quarks and gluons, the question is how a stick-like or string-like object actually emerges from that theory.

One possible answer is via the formation of a color electric flux tube. Imagine that the color electric field running between a static quark and antiquark is, for some reason, squeezed into a cylindrical region, whose cross-sectional area is nearly constant as quark-antiquark separation L increases. In that case the energy stored in the color electric field E_k will grow linearly with quark separation, i.e.

$$\text{Energy} = \sigma L \quad \text{with} \quad \sigma = \int d^2 x_\perp \tfrac{1}{2} E_k^a(x) E_k^a(x) , \tag{3.5}$$

where $E_k^a(x) = F_{0k}^a(x)$, and the integration is over a cross-section of the flux tube. This means that there will be a linearly rising potential energy associated with static sources (the "static quark potential"), and an infinite energy is required to separate these charges an infinite distance. However, this is not quite what happens in real QCD.

In real QCD with light fermions, the linear potential does not extend indefinitely. For sufficiently large quark separations it is energetically favorable to pair-produce a quark and antiquark of masses m_q, and thereby break the flux tube (or QCD string) as indicated in Fig. 3.2. The static quark potential then looks something like the

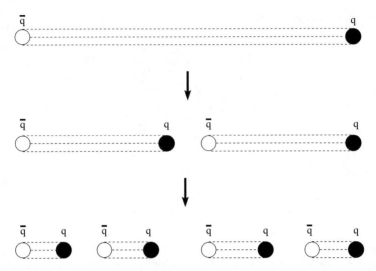

Fig. 3.2 String breaking by quark-antiquark pair production

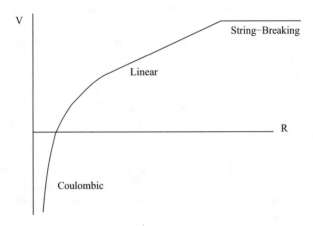

Fig. 3.3 Qualitative features of the static quark potential in QCD, including the effect of quark fields

sketch in Fig. 3.3. At very small distances asymptotic freedom prevails, and the static potential is Coulomb-like. At intermediate distances a flux tube forms, and the potential is linear. At large distances, the color field of the static quarks is screened by the dynamical matter fields. This screening behavior can be demonstrated in strong-coupling lattice gauge theory calculations and, more to the point, it has been seen in lattice Monte Carlo simulations when matter fields are added to the action [2–4].

3.2   The Fradkin-Shenker-Osterwalder-Seiler Theorem

I have already stated, without giving any reasons, that the spectrum of a gauge-Higgs theory, like that of QCD, consists only of color-singlet states. The reasons for thinking this is so were given by Fradkin and Shenker [5], who based their argument on an earlier theorem due to Osterwalder and Seiler [6].

Consider a lattice gauge theory with the Higgs field in the fundamental representation of the $SU(2)$ gauge group, and let the modulus of the Higgs field be fixed. This theory can be realized on the lattice by the action [7]

$$S = -\beta \sum_{plaq} \tfrac{1}{2} \text{Tr}[UUU^\dagger U^\dagger] - \gamma \sum_{x,\mu} \tfrac{1}{2} \text{Tr}[\phi^\dagger(x) U_\mu(x) \phi(x+\hat{\mu})] \,, \qquad (3.6)$$

where $\phi(x)$, like the link variables $U_\mu(x)$, is $SU(2)$ group-valued, and transforms under a local gauge transformation as $\phi(x) \to \phi'(x) = G(x)\phi(x)$. For small values of the Higgs coupling constant γ, the dynamics of the theory resembles QCD, in the sense that there is electric flux tube formation followed by string breaking, and a potential (extracted from Wilson loops) which rises linearly up to the point of string breaking, and then goes flat. At large values of γ, the situation is much more like the weak interactions. There is no electric flux-tube formation, there are only short-range Yukawa-type interactions, and there exist massive vector bosons that might as well, in analogy to the electroweak theory, be identified as W-bosons. Many textbooks on quantum field theory, in defiance of Elitzur's theorem, would call this region a phase of spontaneously broken gauge symmetry. In Chap. 15 I will discuss in what sense this terminology may be true. But for now, I will refer to the large γ area of the phase diagram as a "Higgs-like" region, in contrast to the "confinement-like" region at small γ.

What Osterwalder and Seiler proved was that there is always a path from a point deep in the confinement-like regime, at $\beta, \gamma \ll 1$, to a point deep in the Higgs-like region, at $\beta, \gamma \gg 1$, such that the expectation value of all local, gauge-invariant observables $\langle A(x_1) \rangle$, $\langle B(x_2) \rangle$, $\langle C(x_3) \rangle \dots$ and products $\langle A(x_1)B(x_2)C(x_3)\dots \rangle$ of such observables, vary analytically along the path. What this implies, as emphasized by Fradkin and Shenker, is that there can be no thermodynamic phase transition along the path, and no discontinuous change in the asymptotic particle spectrum. There is no thermodynamic phase boundary which separates the Higgs and confinement-like regions, and by varying the gauge and Higgs couplings the spectrum evolves smoothly from one region to another.

The Osterwalder-Seiler theorem does not cover the entire coupling plane, and it is in principle possible that there are phase transitions somewhere in the coupling constant space, at moderate values of β, γ. It could be, for example, that there is a massless phase somewhere in the phase diagram. The theory has been studied by lattice Monte Carlo techniques, and the situation is indicated schematically in Fig. 3.4. All indications are that the gauge theory is in a single phase, namely the massive phase, at all values of β, γ except $\gamma = 0$, where the theory is in a phase

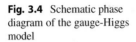

Fig. 3.4 Schematic phase diagram of the gauge-Higgs model

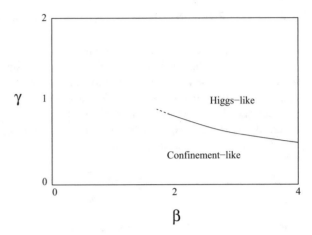

of magnetic disorder. The solid line, from roughly $\beta = 2.0$ to $\beta = 2.7$ indicates a region where certain observables (such as plaquette variables) vary very rapidly with coupling, but still analytically in the infinite volume limit. This is known as a "crossover." The most recent simulations, by Bonati et al. [8], find that this turns into a line of genuine first-order phase transitions at $\beta > 2.7$. But whether the line is a crossover, or a line of genuine first order phase transitions, is not so important. The point is that that there exists a path from any point in the phase diagram, to any other point in the phase diagram, such that all local gauge-invariant observables, and products of such observables, evolve analytically. This means, in particular, that a discontinuous change from a color-singlet spectrum in the confinement-like region, to a color non-singlet spectrum in the Higgs region, is ruled out.

One way of seeing that there are only singlets in the spectrum is to consider the color field emanating from a static color-charged source, in both the confinement-like and Higgs-like regions of the phase diagram. In neither case is the color field detectable far from the source. In the confinement-like region, a scalar particle will bind to the charged source, producing a color singlet which has no associated long-range color field. In the Higgs-like region, gauge forces are short range, and again the color field is undetectable a short distance away from the source. What this means, in terms of the Gauss Law, is that the Higgs field in the vacuum has adjusted itself in the neighborhood of the source to cancel out the static charge. The Fradkin-Shenker work implies that one charge cancellation mechanism evolves continuously into the other, as we move along some path from the confinement-like to the Higgs-like region in the coupling-constant plane.

One might object that gauge particles appear in the spectrum of Higgs theories as massive W-bosons. How can these gauge particles evolve continuously, along some path in the coupling-constant space, from color-singlet states in the confinement-like region? The answer is that the same gauge-invariant operators, which create composite, color singlet vector mesons in the confinement-like region, also create particle states which can be interpreted as massive gauge bosons in the Higgs-like region. An example of such operators in the $SU(2)$ gauge-Higgs theory (3.6) would be

$$\text{Tr}\left[\frac{\sigma_a}{2}\phi^\dagger(x)U_\mu(x)\phi(x+\widehat{\mu})\right] \tag{3.7}$$

where the σ_a denote the Pauli matrices. These color-singlet vector meson operators also create the massive W-bosons in the Higgs-like region, which is made clear by going to a unitary gauge $\phi = \mathbb{1}$, and expanding $U_\mu(x) \approx \mathbb{1} + igA_\mu^a\sigma_a/2$. From this fact it is evident that W-bosons in the Higgs-like region *can* evolve continuously, from states which we would interpret as bound states of scalar particles in the confinement-like region. The fact that particles in the spectrum of the electroweak theory are created by local, gauge invariant operators was first pointed out by Fröhlich, Morchio, and Strocchi [9], and by 't Hooft [10]. In SU(N) gauge theories with $N > 2$, however, there is not a perfect match between the local gauge invariant operators associated with the physical spectrum, and the particle spectrum expected from perturbation theory [11].

3.3 Remnant Gauge Symmetries

Elitzur's theorem, and the Fradkin-Shenker-Osterwalder-Seiler theorem, do some violence to one's preconceptions about gauge-Higgs theories. Consider a Higgs field transforming in the fundamental representation of the gauge group, so in an $SU(N)$ theory the Higgs field $\boldsymbol{\varphi}$ has N components φ^i. Standard introductions to the Higgs mechanism often teach that when the Higgs potential has a certain form, usually a "Mexican-hat", i.e.

$$V(\varphi) = \lambda(\boldsymbol{\varphi} \cdot \boldsymbol{\varphi} - \mu)^2 , \tag{3.8}$$

such that the potential is minimized away from $\boldsymbol{\varphi} = 0$, then the Higgs potential acquires a vacuum expectation value $\langle\boldsymbol{\varphi}\rangle \neq 0$, and the gauge symmetry is spontaneously broken. The fact is, however, that in the absence of gauge fixing we have $\langle\boldsymbol{\varphi}\rangle = 0$ regardless of the form of the Higgs potential. If, on the other hand, we fix to a unitary gauge, where one uses the gauge freedom to rotate the Higgs field into a definite direction in color space, then $\langle\boldsymbol{\varphi}\rangle \neq 0$, also regardless of the Higgs potential. This is just due to quantum fluctuations away from the classical minimum.

Certain gauge choices, however, such as Landau and Coulomb gauge, do not fix all of the gauge symmetry, but leave some global subgroup unfixed. Elitzur's theorem does not apply to these remnant gauge symmetries, and in principle such remnant symmetries *can* break spontaneously. There are a number of suggestions in the literature which argue that the distinction between confining and non-confining gauge theories is related to whether or not some remnant gauge symmetry is, or is not, spontaneously broken. Three examples are the Kugo-Ojima confinement criterion [12, 13], the Coulomb confinement criterion [14], and the Pisa criterion

[15, 16] associated with dual superconductivity (which will be discussed in Chap. 9). A fourth proposal is reserved for Chap. 15.

3.3.1   Landau Gauge and the Kugo-Ojima Criterion

If a gauge field A_μ satisfies the Landau gauge condition $\partial_\mu A_\mu = 0$, then so does the gauge-transformed field $GA_\mu G^\dagger$, where $G(\mathbf{x}, t) = G$ is a spacetime-independent gauge transformation; i.e. a global transformation. There is also, as pointed out by Hata [17], a remnant space-time dependent transformation which preserves the Landau gauge condition:

$$G(x) = \exp\left[i\Lambda^a(\varepsilon; x)\tfrac{1}{2}\sigma_a\right]$$

$$\Lambda^a(\varepsilon; x) = \varepsilon^a_\mu x^\mu - g\frac{1}{\partial^2}(A_\mu \times \varepsilon_\mu)^a + O\left(g^2\right), \tag{3.9}$$

where g is the gauge coupling. This transformation also counts as global, since it depends on a finite number of parameters ε^a_μ, and is therefore not subject to the Elitzur theorem. These remnant symmetries exist in any covariant gauge.

Kugo and Ojima introduce the function $u^{ab}(p^2)$, defined by the expression

$$u^{ab}\left(p^2\right)\left(g_{\mu\nu} - \frac{p_\mu p_\nu}{p^2}\right)$$

$$= \int d^4x\, e^{ip(x-y)}\langle 0|T[D_\mu c^a(x)g(A_\nu \times \overline{c})^b(y)]|0\rangle, \tag{3.10}$$

where $c, \overline{c}$ are ghost and antighost fields required, in the Faddeev-Popov approach, for quantization in any covariant gauge. They then show that the expectation value of color charge in any physical state vanishes

$$\langle \text{phys} | Q^a | \text{phys} \rangle = 0, \tag{3.11}$$

providing that (1) remnant symmetry with respect to spacetime-independent gauge transformations $G(\mathbf{x}, t) = G$ is unbroken; and (2) the following condition is satisfied:

$$u^{ab}(0) = -\delta^{ab}. \tag{3.12}$$

This latter condition is known as the Kugo-Ojima confinement criterion, and it implies that the ghost propagator is more singular, and the gluon propagator less singular, than a simple pole at $p^2 = 0$ [13]. A number of efforts have focussed on verifying this condition (or its corollaries) both analytically [18] and numerically [19].

It turns out that the Kugo-Ojima condition (3.12) is itself tied to the unbroken realization of remnant gauge symmetry in covariant gauges. It was shown by Hata [17] (see also Kugo [13]) that the condition (3.12) is a necessary (and probably sufficient) condition for the unbroken realization of the residual spacetime-dependent symmetry (3.9), while an unbroken, spacetime independent symmetry is required, in *addition* to (3.12), for the vanishing of $\langle \psi | Q^a | \psi \rangle$ in physical states.

Thus the Kugo-Ojima scenario requires that the full remnant gauge symmetry in Landau gauge, i.e. both the spacetime dependent and the spacetime independent residual gauge symmetries, must be unbroken. Both of these symmetries are necessarily broken if a Higgs field acquires an expectation value in a covariant gauge, so in the Kugo-Ojima scheme the confinement phase is a phase of unbroken remnant gauge symmetry.

3.3.2 Coulomb Confinement

The Coulomb gauge condition $\nabla \cdot A = 0$ is preserved by time-dependent, but space-independent gauge transformations $G(\mathbf{x}, t) = G(t)$. This is a remnant symmetry which is local in time, but global in space, and if we restrict ourselves to a single time-slice, or a finite set of time-slices, then in principle this symmetry can break spontaneously.[1] On the lattice, Coulomb gauge is equivalent to maximizing, on each time slice, the quantity

$$R(t) = \sum_{\mathbf{x}} \sum_{k=1}^{3} \mathrm{Tr}[U_k(\mathbf{x}, t)] \,, \qquad (3.13)$$

and this quantity is obviously undisturbed by $U_k(\mathbf{x}, t) \rightarrow G(t)U_k(\mathbf{x}, t)G^\dagger(t)$. On the other hand, for timelike links $U_0(\mathbf{x}, t) \rightarrow G(t)U_0(\mathbf{x}, t)G^\dagger(t+1)$, so if the remnant symmetry is unbroken at fixed times t and $t + 1$, then necessarily

$$\langle \mathrm{Tr}[U_0(\mathbf{x}, t)] \rangle = 0 \qquad (3.14)$$

because the left hand side is not invariant with respect to the remnant symmetry. It will be shown in Chap. 9 that (3.14) implies that the energy of the Coulomb field associated with a static, isolated color charge is infinite, even when the usual ultraviolet divergence is regulated with a lattice cutoff. For this reason, Marinari, Paciello, Parisi and Taglienti [14] (see also [20]) suggested that a possible criterion for confinement is that the remnant gauge symmetry in Coulomb gauge is unbroken. In principle the criterion can work even in the presence of string-breaking matter

---

[1]What this means is that the order parameter for the transition can be non-zero in the infinite 3-volume limit at any fixed time, but have different values at different times, averaging to zero if we average over all times. Thus the symmetry can be spontaneously broken on any single timeslice, or finite set of timeslices, without violating the Elitzur theorem.

fields such as quarks, because the Coulomb field is not necessarily the actual gauge field surrounding a color charge, and the color Coulomb potential is only an upper bound on the static quark potential.

3.3.3   Remnant Symmetry Breaking

An apparent difficulty with confinement criteria which rely on the broken or unbroken status of a remnant gauge symmetry is that, first of all, different remnant symmetries break at different couplings (so which is the "right" remnant symmetry to consider?), and secondly, they break in the absence of a thermodynamic phase transition. The situation for the gauge-Higgs model of (3.6) is shown in Fig. 3.5. We may use, as the order parameter for spontaneous breaking of remnant Landau gauge symmetry, the spatial average of the Higgs field

$$\widetilde{\phi} = \frac{1}{V} \sum_x \phi(x) \,, \tag{3.15}$$

and for spontaneous breaking of remnant symmetry in Coulomb gauge, the spatial average of timelike links on any given timeslice

$$\widetilde{U}(t) = \frac{1}{V_3} \sum_{\mathbf{x}} U_0(\mathbf{x}, t) \,. \tag{3.16}$$

The test is whether or not the expectation values of the magnitudes of $\widetilde{\phi}$ and $\widetilde{U}(t)$ extrapolate to zero as the 4-volume V, and 3-volume V_3, respectively, are taken to infinity.

The result of a lattice Monte Carlo simulation [21] is shown in Fig. 3.5. What we see is that the Coulomb and Landau gauge remnant symmetries break along lines which coincide along the crossover, but diverge from one another at lower β where there is, in fact, no line of thermodynamic transitions. This calls into question both the Kugo-Ojima criterion, and the Coulomb confinement criterion. A third example (the Pisa criterion [15, 16]), which identifies confinement as the broken phase of a remnant "dual" $U(1)$ gauge symmetry, was studied in Ref. [22]. It turns out that this criterion also predicts transitions between confining and non-confining phases.

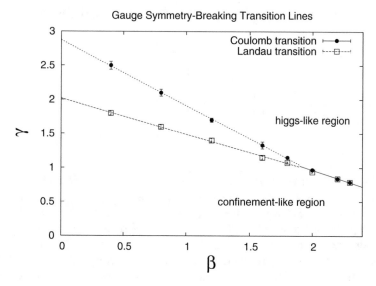

Fig. 3.5 The location of remnant global gauge symmetry breaking in Landau and Coulomb gauges, in the $\beta - \gamma$ coupling plane. From [21]

3.4 Center Symmetry

In the Ising model example, the distinction between the ordered and disordered phases corresponded to whether or not a certain global symmetry of the theory is spontaneously broken. But in the gauge-Higgs theory we have discussed, the distinction seems to be ambiguous (although we will revisit this quesiton in Chap. 15). If we classify the theory based on the behavior of very large Wilson loops, we find that the theory is *almost* everywhere in a massive phase, characterized by a perimeter-law falloff. The exception is along the line at $\gamma = 0$. As this line is approached, string-breaking occurs at ever larger charge separations. Precisely at $\gamma = 0$ the Higgs field decouples from the gauge field, string breaking disappears entirely and (it is believed) Wilson loops have an area-law falloff at arbitrarily large scales. In other words, at $\gamma = 0$ the system is in a phase of magnetic disorder. The same thing happens in a gauge theory like QCD, if the quark masses are taken to infinity. It is then worth asking: is there some global symmetry which distinguishes the massive phase from the magnetically disordered phase?

There is. It is known as *center symmetry*, and, although a global symmetry, it is associated with the center of the gauge group.

We will need a few basic facts from group theory. First of all, the center of a group consists of those group elements which commute with all elements of the group. For an $SU(N)$ gauge theory, this is the set of all group elements proportional to the identity:

$$z_n \mathbb{1}_N = \exp\left(\frac{2\pi i n}{N}\right) \begin{bmatrix} 1 & & & & \\ & 1 & & & \\ & & \cdot & & \\ & & & \cdot & \\ & & & & \cdot \\ & & & & & 1 \end{bmatrix} \qquad (n = 0, 1, 2, ..., N-1) . \qquad (3.17)$$

The $\{z_n = \exp[2\pi i n/N]\}$ are the elements of the discrete abelian subgroup Z_N, and the set $\{z_n \mathbb{1}_N\}$ is the Z_N subgroup of $SU(N)$. While there are an infinite number of representations of $SU(N)$, there are only N representations of Z_N. This means that every representation of $SU(N)$ falls into one of N subsets (known as *class* or *N-ality*), depending on the representation of the Z_N subgroup. The N-ality k of a given representation is given by the number of boxes in the corresponding Young Tableau, mod N. If $M[g]$ is the matrix representation of group element g in a representation of N-ality k, and $z \in Z_N$ is an element in the center, then

$$M[zg] = z^k M[g] . \qquad (3.18)$$

Now consider the lattice action (2.24) of a pure gauge theory, i.e. a gauge theory with no matter fields. It is easy to see that this action is invariant under the following transformation: Pick any fixed time $t = t_0$ and multiply the timelike links on that timeslice by an element of the center subgroup, i.e.

$$U_0(\mathbf{x}, t_0) \rightarrow z U_0(\mathbf{x}, t_0) \quad \text{all } \mathbf{x} , \text{ fixed } t_0 , \qquad (3.19)$$

with $z \in Z_N$ if the gauge group is $SU(N)$; all other link variables are unchanged. The transformation is indicated in Fig. 3.6; transformations defined at other fixed times are equivalent up to a gauge transformation. The only elements of the lattice action which might be affected are the timelike plaquettes containing the links $U_0(\mathbf{x}, t_0)$ but in fact these are unchanged:

$$U_i(\mathbf{x}, t_0) z U_0(\mathbf{x} + \hat{i}, t_0) U_i^\dagger(\mathbf{x}, t_0 + 1) U_0^\dagger(\mathbf{x}, t_0) z^{-1}$$
$$= U_i(\mathbf{x}, t_0) U_0(\mathbf{x} + \hat{i}, t_0) U_i^\dagger(\mathbf{x}, t_0 + 1) U_0^\dagger(\mathbf{x}, t_0) \qquad (3.20)$$

because of the fact that the center elements commute with all elements of the group, and with any link variable in particular.

The reason that center symmetry is so important is that if this symmetry

- is broken explicitly, e.g. by matter fields in the fundamental representation of the gauge group, or

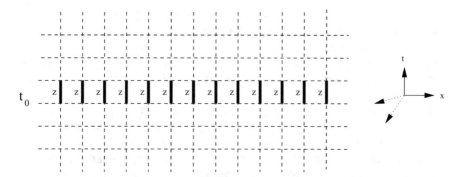

Fig. 3.6 The global center transformation. Each of the indicated links in the time direction, on the timeslice $t = t_0$, is multiplied by the same element z of the center subgroup. The lattice action of a pure gauge theory is left unchanged by this operation

- is broken spontaneously, which happens at high temperature, and *may* happen if there are matter fields in the adjoint representation of the gauge group, or
- is trivial, i.e. the center subgroup consists only of the unit element, as in the exceptional group G_2,

then the static quark potential cannot be linear asymptotically; it must become flat.

Let's see why matter fields in the fundamental representation, or in any representation of the gauge group of N-ality $k \neq 0$, must break the center symmetry explicitly. A matter field in color representation r couples to the gauge field via a term in the action

$$\gamma \sum_{x,\mu} \phi^\dagger(x) U_\mu^{(r)}(x) \phi(x + \hat{\mu}) + \text{c.c.} \,, \tag{3.21}$$

where the superscript r means that the link variables are in the r-representation. Under the global center symmetry transformation

$$\phi^\dagger(\mathbf{x}, t_0) U_0^{(r)}(\mathbf{x}, t_0) \phi(\mathbf{x}, t_0 + 1) \rightarrow z^k \phi^\dagger(\mathbf{x}, t_0) U_0^{(r)}(\mathbf{x}, t_0) \phi(\mathbf{x}, t_0 + 1) \,. \tag{3.22}$$

So if the N-ality k of the representation of the matter field is non-zero, the matter action breaks the global Z_N center symmetry. But matter fields of N-ality $k \neq 0$ are exactly the fields associated with string-breaking. Suppose we have two static color charges, in the fundamental representation of the gauge group. In order to break the electric flux tube which forms between these sources, by the process pictured in Fig. 3.2, there must be dynamical particles which can bind to the sources and produce a color singlet. This is not possible if the particles are in a color representation with N-ality $k = 0$ (such as gluons, or any matter fields in an adjoint representation). Gluons can screen color in the sense that they can bind to a source of N-ality $k \neq 0$, and produce a particle with color in a lower dimensional representation of the gauge group with the *same* N-ality k, but it is not possible for the bound state to change the N-ality of the source. Particles with color charge in the fundamental representation, however, with $k = 1$, can bind and form a color singlet with any color-charged source. So if Z_N center symmetry is completely broken by $k = 1$ matter fields, then the static potential between any two sources must ultimately become flat.

In the case of group such as G_2, in which the center subgroup is only the unit element, gluons can bind with any color charge to form a singlet. So in this theory also, the static potential between any two sources is flat at large distances.

We will see in the next chapter that, if center symmetry is unbroken, then the energy of an isolated color charge in the fundamental representation is infinite, and the potential energy of a particle-antiparticle pair rises without limit as the charges separate. Conversely, if the symmetry is spontaneously broken, the energy of an isolated color charge is finite, and a particle-antiparticle pair *can* be separated.

In short, center symmetry is associated with an infinitely rising static quark potential, and is therefore the global symmetry associated with the magnetically disordered phase of a gauge theory. When the symmetry is broken in some way, then the gauge theory exists in some other phase.

Now it is time to return to the semantic question that opened this chapter. What is "confinement?" We have seen that "color confinement" exists in a Higgs theory; it is not necessarily associated with a linear potential, flux tube formation, and Regge trajectories. It is perfectly all right to define confinement as color confinement, providing it is understood that such a term does not necessarily refer to the phenomena we are interested in. I prefer to reserve the term "confinement" to refer to a particular phase of gauge theories, namely, the magnetically disordered phase. Confinement in this sense is not the same thing as color confinement (although that is implied), nor is it the same thing as a mass gap.[2] This terminology has a price; it means that QCD is a "confinement-like" theory, i.e. it resembles a magnetically disordered theory only below some string-breaking length scale, which goes off to infinity as the quark masses are taken to infinity. But string-breaking is easy to understand; the hard problem is to understand the origins of the linear potential. For this reason it makes sense to concentrate on magnetic disorder. The confinement problem is solved if we can understand why non-abelian gauge theories with a non-trivial center symmetry can only exist, in $D \leq 4$ dimensions, in the magnetically disordered phase.

This still leaves open the question of how to define the word "confinement" with precision, when there can exist string breaking via particles which transform in the fundamental representation of the gauge group. I will return to this question at the end of this book.

---

[2]The Clay Mathematics Institute offers a large cash prize (http://www.claymath.org/millennium/ Yang-Mills_Theory) for proving that Yang-Mills theory has a mass gap, which would be true in any gauge theory which is in either a magnetically disordered phase, as in pure Yang-Mills theory, or a massive phase, as in Higgs theories and in real QCD.

References

1. G. Bali, QCD forces and heavy quark bound states. Phys. Rep. **343**, 1 (2001). arXiv: hep-ph/0001312
2. O. Philipsen, H. Wittig, String breaking in non-Abelian gauge theories with fundamental matter fields. Phys. Rev. Lett. **81**, 4056 (1998). Erratum-ibid. **83**, 2684 (1999). arXiv:hep-lat/9807020
3. F. Knechtli, String breaking and lines of constant physics in the SU(2) Higgs model. Nucl. Phys. Proc. Suppl. **83**, 673 (2000). arXiv:hep-lat/9909164
4. C.E. Detar, U.M. Heller, P. Lacock, First signs for string breaking in two-flavor QCD. Nucl. Phys. Proc. Suppl. **83**, 310 (2000). arXiv:hep-lat/9909078
5. E. Fradkin, S. Shenker, Phase diagrams of lattice gauge theories with Higgs fields. Phys. Rev. D **19**, 3682 (1979)
6. K. Osterwalder, E. Seiler, Gauge field theories on the lattice. Ann. Phys. **110**, 140 (1978)
7. C. Lang, C. Rebbi, M. Virasoro, The phase structure of a non-Abelian gauge Higgs system. Phys. Lett. **104B**, 294 (1981)
8. C. Bonati, G. Cossu, M. D'Elia, A. Di Giacomo, Phase diagram of the lattice SU(2) Higgs model. Nucl. Phys. B **828**, 390 (2010). arXiv:0911.1721 [hep-lat]
9. J. Frohlich, G. Morchio, F. Strocchi, Higgs phenomenon without symmetry breaking order parameter. Nucl. Phys. B **190**, 553 (1981)
10. G. 't Hooft, Which topological features of a gauge theory can be responsible for permanent confinement?. NATO Sci. Ser. B **59**, 117 (1980)
11. A. Maas, R. Sondenheimer, P. Törek, On the observable spectrum of theories with a Brout?Englert?Higgs effect. Ann. Phys. **402**, 18 (2019). https://doi.org/10.1016/j.aop.2019.01.010. arXiv:1709.07477 [hep-ph]
12. T. Kugo, I. Ojima, Local covariant operator formalism of nonabelian gauge theories and quark confinement problem. Prog. Theor. Phys. Suppl. **66**, 1 (1979)
13. T. Kugo, The Universal Renormalization Factors Z(1) / Z(3) and Color Confinement Condition in nonAbelian Gauge Theory. arXiv:hep-th/9511033
14. E. Marinari, M.L. Paciello, G. Parisi, B. Taglienti, The string tension in gauge theories: a Suggestion for a new measurement method. Phys. Lett. B **298**, 400 (1993). arXiv:hep-lat/9210021
15. L. Del Debbio, A. Di Giacomo, G. Paffuti, Detecting dual superconductivity in the ground state of gauge theory. Phys. Lett. B **349**, 513 (1995). arXiv:hep-lat/9403013
16. A. Di Giacomo, B. Lucini, L. Montesi, G. Paffuti, Color confinement and dual superconductivity of the vacuum. 1. Phys. Rev. D **61**, 034503 (2000). arXiv:hep-lat/9906024
17. H. Hata, Restoration of the local gauge symmetry and color confinement in non-Abelian gauge theories. II. Prog. Theor. Phys. **69**, 1524 (1983). Prog. Theor. Phys. **67**, Restoration of the local gauge symmetry and color confinement In non-Abelian gauge theories, 1607 (1982)
18. R. Alkofer, C.S. Fischer, F.J. Llanes-Estrada, K. Schwenzer, The quark-gluon vertex in Landau gauge QCD: its role in dynamical chiral symmetry breaking and quark confinement. Ann. Phys. **324**, 106 (2009). arXiv:0804.3042 [hep-ph]
19. H. Nakajima, S. Furui, A. Yamaguchi, In: *Proceedings of the 30th International Conference on High-Energy Physics (ICHEP 2000)*. Numerical study of the Kugo-Ojima criterion and the Gribov problem in the Landau gauge. arXiv:hep-lat/0007001
20. J. Greensite, S. Olejnik, D. Zwanziger, Coulomb energy, remnant symmetry, and the phases of non-Abelian gauge theories. Phys. Rev. D **69**, 074506 (2004). arXiv:hep-lat/0401003
21. W. Caudy, J. Greensite, On the ambiguity of spontaneously broken gauge symmetry. Phys. Rev. D **78**, 025018 (2008). arXiv:0712.0999 [hep-lat]
22. J. Greensite, B. Lucini, Is confinement a phase of broken dual gauge symmetry? Phys. Rev. D **78**, 085004 (2008). arXiv:0806.2117 [hep-lat]

Order Parameters for Confinement

4

Abstract

The Wilson loop, Polyakov loop, and 't Hooft loop as order parameters for confinement. Wilson loops and the propagation of heavy quark-antiquark pairs, Polyakov loops and finite temperature gauge theories, 't Hooft loops as center vortex creation operators, and the center vortex free energy. The topological linking of center vortices with Wilson loops. The relation between order parameters for confinement, and center symmetry.

There are certain observables whose behavior indicates that a gauge system is, or is not, in a phase of magnetic disorder. These are the Wilson loop, the Polyakov loop, the 't Hooft loop, and the vortex free energy. The Wilson loop has been discussed already, to some extent. The Polyakov loop is a Wilson loop which winds around a finite, periodic lattice in the time direction; it is a crucial probe of center symmetry and the high temperature deconfinement phase transition, in which center symmetry is spontaneously broken. The 't Hooft loop is an operator which creates an object known as a *center vortex*. The center vortex scenario for confinement will be discussed in some detail in Chaps. 6–8.

4.1 The Wilson Loop

The Wilson loop operator [1] can be thought of in two ways:

1. rectangular timelike loops represent the creation, propagation, and destruction of two static sources of quark and antiquark color charge, respectively, located at fixed points in space (cf. Chap. 2);
2. spacelike loops are probes of gauge-field fluctuations in the vacuum state, independent of any external source.

© Springer Nature Switzerland AG 2020

J. Greensite, *An Introduction to the Confinement Problem*, Lecture Notes in Physics 972, https://doi.org/10.1007/978-3-030-51563-8_4

Of course, spacelike and timelike loops are not intrinsically different, but are related to one another by Lorentz (or, in the Euclidean formulation, rotation) invariance. This means that the interaction energy between static sources is related to vacuum fluctuations in the absence of external sources.

Let us again consider the relation of timelike loops to the static quark potential, this time for static charges in any group representation r. The theory consists of an $SU(N)$ lattice gauge field coupled to a single, very massive, scalar field in color group representation r. The action is

$$S = -\frac{\beta}{N} \sum_p \mathrm{ReTr}[U(p)] + S_{matter} , \tag{4.1}$$

where

$$S_{matter} = -\sum_{x,\mu} \left(\phi^\dagger(x) U_\mu^{(r)}(x) \phi(x+\hat{\mu}) + \mathrm{c.c.} \right) + \sum_x (m^2 + 2D)\phi^\dagger(x)\phi(x) , \tag{4.2}$$

and $U_\mu^{(r)}(x)$ denotes the link variable in group representation r. Again let Q be an operator which creates a two-particle color-singlet state, with separation R,

$$Q(t) = \phi^\dagger(0,t) \left\{ \prod_{n=0}^{R-1} U_i^{(r)}(n\hat{i},t) \right\} \phi(R\hat{i},t) . \tag{4.3}$$

The Q operator not only creates the two charged massive particles, but also creates a thin line of color electric flux running between those particles. In general this is not the minimal energy configuration for the color electric field.

We then consider the vacuum expectation value $\langle Q^\dagger(T)Q(0)\rangle$. This expression can be interpreted as creating a massive color charged pair joined by a line of color electric flux, which propagates for some Euclidean time T, and is then annihilated. As $m^2 \to \infty$ string-breaking effects can be neglected, and we can integrate out the scalar field

$$\langle Q^\dagger(T)Q(0)\rangle \sim m^{-4(T+1)} \frac{1}{Z} \int DU \ \chi_r[U(R,T)]e^{-S_U}$$

$$\sim m^{-4(T+1)} W_r(R,T) . \tag{4.4}$$

Note that $W_r(R,T)$ is computed in the pure gauge theory, with no matter fields. Here $\chi_r[g]$ is the *group character* of group element g in representation r, S_U is the action without the S_{matter} term, $U(R,T)$ is the *holonomy*, i.e. the ordered product of link variables around the loop, and $W_r(R,T)$ is the expectation value

$$W_r(R,T) = \left\langle \chi_r[U(R,T)] \right\rangle . \tag{4.5}$$

A "holonomy" is just a Wilson loop before taking the trace; a "group character" is the trace of a group element in the given representation. In the continuum, the holonomy $U(C)$ is the path-ordered Wilson loop

$$U(C) = P \exp\left[ig \oint_C dx^\mu \, A_\mu(x) \right], \tag{4.6}$$

where the path-ordering symbol P means that, in the power-series expansion of the exponential, a product of $A_\mu(x)$ operators at different points on the loop are ordered according to where they appear on the loop relative to some beginning point on the loop.

Then we have, by the same logic as (2.33)

$$W_r(R, T) \propto \sum_n |c_n|^2 e^{-\Delta E_n T}. \tag{4.7}$$

where the sum runs over states containing two charged particles separated by a distance R, and ΔE_n is the energy of the n-th state above the vacuum energy. The minimal energy state of the color field joining the charged particles is singled out in the $T \to \infty$ limit, and

$$V_r(R) = -\lim_{T \to \infty} \log\left[\frac{W_r(R, T+1)}{W_r(R, T)} \right] \tag{4.8}$$

is the static quark potential for color sources in group representation r. Again, this is the potential for massive color charged sources. We assume the absence of light matter fields which could, by pair production, screen the charge of the massive sources. The confinement problem is to show that

$$V_r(R) \sim \sigma_r R \tag{4.9}$$

at large R, for non-zero N-ality representations, or more generally an area-law falloff for Wilson loops

$$W_r(C) \sim \exp[-\sigma_r \, \text{Area}(C)] \tag{4.10}$$

for contours C enclosing a large minimal area.

The relationship of the Wilson loop to the potential between massive, color-charged particles, and the fact that the timelike lines of a loop can be interpreted as the worldlines of static particles, sometimes obscures the fact that the Wilson loop is an observable which measures the vacuum fluctuations of gauge fields, and that this observable makes sense even if there are no matter particles whatever in the theory. This is particularly clear for spacelike loops C, contained entirely in some spatial volume at constant time. In that case

$$W_r(C) = \langle \Psi_0 | \chi_r[U(C)] | \Psi_0 \rangle, \tag{4.11}$$

where Ψ_0 is the ground state of the gauge-field Hamiltonian, and there are no charged sources whatever. In fact, in relating the Wilson loop to the static potential, we assumed that the mass of the matter field was large but finite. On the other hand, if the mass is finite there must be string-breaking at some distance scale. To avoid string-breaking it is necessary to take the $m \rightarrow \infty$ limit, in which case the correlator in Eq. (4.4) is actually zero. In extracting the Wilson loop we have factored out the mass term, retaining the holonomy that depends only on the gauge field, and whose expectation value makes sense in the $m \rightarrow \infty$ limit.

It should also be noted that the expectation value of a Wilson loop is always defined with an implicit (usually lattice) ultraviolet regularization. In the continuum limit, the gluon propagator is singular at short distances, and this leads to a perimeter contribution to the expectation value with a singular coefficient. For timelike loops, this singularity can be viewed as a divergent self-energy of the charged particles and antiparticles running along the loop. To really make sense of Wilson loops in the continuum, some "smearing" of the loop, i.e. a superposition of many nearby loops, is required (see, e.g. [2]). The same is true for the Polyakov lines discussed in the next section.

4.2   The Polyakov Loop

Consider an $SU(N)$ gauge theory with either no matter fields, or only N-ality=0 matter fields, formulated on a lattice which is finite and periodic in the time direction. As we have seen, the action is invariant under the global Z_N group of center symmetry transformations

$$U_0(\mathbf{x}, t_0) \rightarrow z U_0(\mathbf{x}, t_0) \ , \quad z \in Z_N \ , \quad \text{all } \mathbf{x} \ , \tag{4.12}$$

implemented on all timelike link variables at any fixed time t_0. This transformation, which we have already encountered in Sect. 3.4, does not change the action or plaquette variables or Wilson loops. But there are certain gauge-invariant observables which *are* affected. Consider the trace of a loop holonomy that winds once around the lattice in the periodic time direction, i.e.

$$P(\mathbf{x}) = \text{Tr}\,[U_0(\mathbf{x}, 1) U_0(\mathbf{x}, 2) \ldots U_0(\mathbf{x}, L_t)] \ , \tag{4.13}$$

where L_t is the extension of the lattice in the time direction. This observable is known as a *Polyakov loop* [3]. Under a center transformation $U_0(x, t_0) \rightarrow z U_0(x, t_0)$, we find

$$P(\mathbf{x}) \rightarrow z P(\mathbf{x}) \ . \tag{4.14}$$

Since the Polyakov loop is not invariant under the global center transformation, its expectation value serves as an order parameter for the spontaneous breaking of the symmetry. There are two possibilities:

$$\langle P(\mathbf{x}) \rangle = \begin{cases} 0 & \text{unbroken } Z_N \text{ symmetry phase} \\ \text{non-zero} & \text{broken } Z_N \text{ symmetry phase} \end{cases} . \tag{4.15}$$

Which of these possibilities is realized by the theory has a lot to do with confinement, because Polyakov loops are, in fact, order parameters for the confinement property. In order to understand this important point, a brief digression into the thermodynamics of quantum field theory is called for.

On the lattice, finite temperature is represented by time-asymmetric lattices, with temperature proportional to $1/L_t$, where L_t is the lattice length in the time direction. The argument for this is straightforward. In quantum statistical mechanics, the partition function is defined to be

$$Z = \text{Tr} e^{-\beta H}$$
$$= \sum_n \langle n | e^{-\beta H} | n \rangle , \tag{4.16}$$

where, in the standard notation, β is not the lattice coupling but rather $\beta = 1/T$, where T is the temperature.[1] The sum is over any complete orthonormal set of states. Now in non-relativistic quantum mechanics, the relationship between the Hamiltonian and path-integral formulations is given by

$$\langle x_f | e^{-iH(t_f - t_0)} | x_i \rangle = \int Dx(t_0 < t < t_f) e^{iS} , \tag{4.17}$$

with the integral over paths $x(t)$ subject to boundary conditions $x(t_0) = x_i$, $x(t_f) = x_f$. Likewise in field theory, after a Wick rotation to Euclidean time

$$\langle \phi_f | e^{-H\beta} | \phi_i \rangle = \int D\phi(\mathbf{x}, 0 < t < \beta) e^{-S} , \tag{4.18}$$

with boundary conditions $\phi(\mathbf{x}, 0) = \phi_i(\mathbf{x})$, $\phi(\mathbf{x}, \beta) = \phi_f(\mathbf{y})$. Here $|\phi(\mathbf{x})\rangle$ represents an eigenstate of the field operator; i.e. it is a state in which the quantum field has a definite value at each point. The set of all such states is a complete set of states in the field theory. This means that the partition function at finite temperature can be expressed as a path integral

$$Z = \int D\phi(\mathbf{x}, 0) D\phi(\mathbf{x}, \beta) \, \delta[\phi(\mathbf{x}, 0) - \phi(\mathbf{x}, \beta)] \int D\phi(x, 0 < t < \beta) e^{-S}$$
$$= \int D\phi(x, 0 \le t \le \beta) \, \delta[\phi(\mathbf{x}, 0) - \phi(\mathbf{x}, \beta)] e^{-S} , \tag{4.19}$$

---

[1] The notation, while standard, is potentially confusing, because β is also used for lattice coupling, and T for time; so I emphasize again that in this section T stands for temperature, not time. In fact, as we will see, it is β which represents a time extension.

where the delta function simply imposes periodic boundary conditions in the time direction. We can also just write

$$Z = \int D\phi(x, 0 \leq t < \beta) \, e^{-S} \tag{4.20}$$

with the understanding that periodic boundary conditions in time are imposed.

Thermal averages, in quantum statistical mechanics, are defined as

$$\langle Q \rangle = \frac{1}{Z} \text{Tr} e^{-\beta H} Q , \tag{4.21}$$

and by the same reasoning as above, this quantity is given by

$$\langle Q \rangle = \frac{1}{Z} \int D\phi(x, 0 \leq t < \beta) \, Q[\phi(\mathbf{x}, 0)] e^{-S} \tag{4.22}$$

with periodic boundary conditions in the time direction. After lattice regularization, $\beta = L_t a$, where a is the lattice spacing. Then the temperature is related to the extension of the periodic lattice in the time direction via $T = 1/(L_t a)$.

A quantity of particular interest in non-abelian gauge theory is the difference F_q between the free energy of a gauge field theory containing an isolated, static quark, and the free energy of a pure gauge system. Let Tr_q denote the trace over states with a single static color charge, due to a massive quark, at some point $\mathbf{x}$. Then

$$\begin{aligned}
e^{-\beta F_q} &\equiv \frac{Z_q}{Z} \\
&= \frac{\text{Tr}_q e^{-\beta H}}{\text{Tr} e^{-\beta H}} \\
&= \frac{\sum_n \langle \Psi_n(1 \text{ quark})|e^{-\beta H}|\Psi_n(1 \text{ quark})\rangle}{\sum_n \langle \Psi_n(0 \text{ quark})|e^{-\beta H}|\Psi_n(0 \text{ quark})\rangle} .
\end{aligned} \tag{4.23}$$

Passing to the path-integral representation for thermal averages, the 1-quark state is obtained by creating a massive quark at point $\mathbf{x}$ and time $t = 0$, propagation of the static quark to point $\mathbf{x}, t = \beta$, and annihilation at $t = \beta$. Integrating over the massive quark field at times $0 \leq t < \beta$ produces a Wilson line running between $(\mathbf{x}, 0)$ and $(\mathbf{x}, \beta)$, and the imposition of periodic boundary conditions results in a closed Wilson loop running through the periodic lattice. Discarding an overall constant proportional to $m^{-\beta}$, the final result is that

$$\begin{aligned}
e^{-\beta F_q} &\propto \frac{1}{Z} \int DU_\mu \, \text{Tr}\left[U_0(\mathbf{x}, 1)U_0(\mathbf{x}, 2) \ldots U_0(\mathbf{x}, L_t)\right] e^{-S} \\
&= \langle P(\mathbf{x}) \rangle ,
\end{aligned} \tag{4.24}$$

with periodic boundary conditions in the time direction understood.

This is the crucial result: the free energy F_q of a quark on the lattice, at temperature $T = 1/(L_t a)$, is infinite if (and only if) the Polyakov loop expectation value is zero, and the center symmetry is unbroken. We can therefore identify the unbroken Z_N center symmetry phase (the "disordered phase") with the confinement phase, and the Polyakov line is an order parameter for confinement. In general, center symmetry is broken spontaneously at high temperatures, so that $\langle P(\mathbf{x}) \rangle \neq 0$ and isolated quarks are finite energy states. This means that quarks are unconfined, in the sense that there exist isolated, finite energy states with the flavor quantum numbers of quarks. The broken center symmetry phase is known as the *deconfined phase*, separated from the confined phase (in a center-symmetric theory) by a deconfinement phase transition which occurs at some critical temperature.[2]

4.3 The 't Hooft Loop

The center symmetry transformation (4.12) can be generalized; it is a special case of a so-called "singular gauge transformation." Consider a transformation

$$U_0(x, t) \rightarrow G(x, t)U_0(x, t)G^\dagger(x, t+1) \tag{4.25}$$

in which $G(x, t)$ is non-periodic in the periodic time direction, or more precisely, is periodic only up to a center transformation

$$G(x, L_t + 1) = z^* G(x, 1) . \tag{4.26}$$

This again transforms Polyakov lines $P(x) \rightarrow z P(x)$, but plaquettes and ordinary Wilson loops are not affected. In the continuum, the singular gauge transformation amounts to transforming the gauge field by the usual formula

$$A_\mu(x) \rightarrow G(x)A_\mu(x)G^\dagger(x) - \frac{i}{g}G(x)\partial_\mu G^\dagger(x) , \tag{4.27}$$

except that at $t = L_t$ and $\mu = 0$ we drop the second term (which would be a delta-function). Because of this, a "singular gauge transformation" is not a true gauge transformation.

Instead of gauge transformations which are discontinuous on loops which wind around the periodic lattice, we could also consider transformations which are discontinuous on other sets of loops. Unlike the center symmetry transformations

---

[2]We note again that an order parameter for the spontaneous breaking of a global symmetry, such as $\langle P(\mathbf{x}) \rangle$, is only non-zero, strictly speaking, in the infinite volume limit. In practice what is done is to compute, on each lattice, the magnitude of the sum of Polyakov loops of the lattice, divided by the number of loops (cf. (6.65) below). The expectation value of this quantity, which is guaranteed to be positive, is then extrapolated to the infinite volume limit, and if the limit is vanishing, the symmetry is unbroken.

we have already discussed, these singular gauge transformations do not leave the action invariant; rather, they create a singular loop of magnetic flux known as a "*thin center vortex.*" There is, in fact, a familiar example of such a vortex creation operator in classical electrodynamics: the exterior field of a solenoid is the result of a singular gauge transformation applied to $A_\mu = 0$.

So let's start with electrodynamics. The Wilson loop holonomy (which is an element in the $U(1)$ gauge group) is

$$U(C) = \exp\left[ie \oint_C dx^\mu A_\mu(x)\right] = e^{ie\Phi_B} , \qquad (4.28)$$

where we take C to be a closed spacelike loop, and Φ_B is the magnetic flux through the loop. If the loop C winds around the exterior of a solenoid, it is certainly possible to have $B = 0$ everywhere along the loop, yet $\Phi_B \neq 0$. In the case of a solenoid of radius R, oriented along the z-axis, we may have

$$A_\theta = \frac{\Phi_B}{2\pi r}\hat{\theta} \quad (r > R) , \qquad (4.29)$$

which is obtained from a singular gauge transformation of $A_\mu = 0$, with

$$G(r, \theta, z, t) = \exp\left[-ie\Phi_B \frac{\theta}{2\pi}\right] . \qquad (4.30)$$

Note the discontinuity

$$G(r, \theta = 2\pi, z, t) = e^{-i\Phi_B} G(r, \theta = 0, z, t) . \qquad (4.31)$$

The gauge transformation is discontinuous on a surface (defined by $y = 0, x > 0$, all z) with a boundary at the z-axis. In obtaining A from the gauge transformation g, we drop the delta-function that would arise in A_θ due to the discontinuity of G on the $y = 0, x > 0$ surface, and which would have to be included if this were a true gauge transformation.

Suppose the gauge transformation $G(r, \theta, z, t)$ has the discontinuity (4.31) for any $r > 0$. Then the effect of the singular gauge transformation on any loop winding once around the z-axis is

$$U(C) \rightarrow e^{\pm ie\Phi_B} U(C) , \qquad (4.32)$$

where the sign in the exponent depends on whether the loop runs clockwise or counterclockwise around the z-axis. From (4.32) and (2.44) it is clear that the singular gauge transformation has created a line of magnetic flux along the z-axis.

We can easily generalize this construction, to obtain an operator which creates a line of magnetic flux running along a closed curve C'. Let S be some surface bounded by the loop C', and let C be a loop which is topologically linked to C',

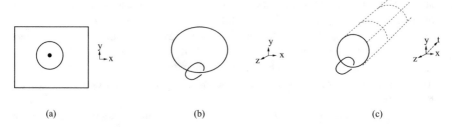

(a)   (b)   (c)

Fig. 4.1 A loop can be topologically linked to: (**a**) a point, in $D = 2$ dimensions; (**b**) another loop, in $D = 3$ dimensions; (**c**) a surface, in $D = 4$ dimensions

parametrized by $\mathbf{x}(\tau)$, where τ runs from $\tau = 0$ to $\tau = 1$, and $\mathbf{x}(0) = \mathbf{x}(1)$ lie on S. Then we consider a singular gauge transformation $G(\mathbf{x})$ which is discontinuous on S; i.e. for any loop C linked to C'

$$G(\mathbf{x}(1)) = e^{-ie\Phi_B} G(\mathbf{x}(0)) . \tag{4.33}$$

Once again, for such loops the singular gauge transformation changes the loop holonomy according to (4.32), so we must conclude that a singular magnetic field has been created along the loop C'.

"Linking" is a topological concept: two loops are linked if it is impossible to translate one loop an arbitrary distance from the other loop without the two objects actually crossing one another. Loops can be linked in three dimensions, but not in four. Suppose two loops C_1 and C_2 appear to be linked at a given time t. The loops can be separated an arbitrary distance, without crossing, by simply translating one of them in time. So the sorts of objects that can be linked depends on spacetime dimension, as indicated in Fig. 4.1, and the classification is as follows:

1. In $D = 2$ dimensions, loops can link to points. A singular gauge transformation is discontinuous along a line ending at a point, and any loop linked to the point is changed by the transformation according to (4.32). The transformation creates a singular magnetic field (a thin vortex) at a point.
2. In $D = 3$ dimensions, loops link to loops. A singular gauge transformation, discontinuous on a surface S bounded by loop C', creates a singular magnetic field (thin vortex) along loop C'. Again, a holonomy along loop C, linked to C', is changed according to (4.32).
3. In $D = 4$ dimensions loops link to surfaces. A singular gauge transformation, discontinuous on a 3-volume V_3 which is bounded by a surface S, creates a surface of magnetic flux on S. A holonomy along loop C, linked to the surface S, is changed according to (4.32).

The general case, of course, follows by induction: In D dimensions, loops link to $D - 2$ dimensional hypersurfaces. Now in $D = 2$ dimensions a loop can wind more than once around a point; the number of times that the loop goes around the point is known as the *winding number*. In $D = 3$ dimensions two loops can obviously wind more than once around each other, and the topological invariant which describes this intertwining is known as the *linking number*. The concept of linking number generalizes to describe the topological connectivity of a surface and a loop in $D = 4$ dimensions, but for now it will be sufficient to consider only connections with linking number = 1.

Returning to $U(1)$ gauge theory, we see that a singular gauge transformation which is discontinuous on a region of dimension $D - 1$, bounded by a hypersurface of dimension $D - 2$, creates a singular magnetic field (a thin vortex) on the $D - 2$ dimensional hypersurface. The $D - 1$ dimensional region of discontinuity is known as a Dirac line, surface, 3-volume, etc., depending on the dimension D. The same procedure can be applied to define a magnetic vortex creation operator in $SU(N)$ theories, but with one importance difference as compared to abelian gauge theories. In an abelian gauge theory, the discontinuity (4.33) consists of multiplication by *any* element $\exp[ie\Phi_B]$ of the abelian gauge group. In a non-abelian theory, if the singular gauge transformation is applied to a non-zero field configuration, this transformation would not only change the action at the location of the $D - 2$ dimensional thin vortex, but would also affect the action all along the $D - 1$ dimensional volume of discontinuity. To avoid changing the action on the discontinuity volume, in the same way that this is avoided in (3.20), and create *only* a magnetic vortex on the $D - 2$ dimensional surface, the discontinuity must be restricted to multiplication by a center element; i.e.

$$G(\mathbf{x}(1)) = zG(\mathbf{x}(0)) \quad , \quad z \in Z_N \tag{4.34}$$

for $SU(N)$ gauge theories. Then any loop holonomy $U(C)$ which is topologically linked to the vortex, with linking number = 1, will transform as

$$U(C) \rightarrow z^*U(C) . \tag{4.35}$$

For linking number l, $U(C) \rightarrow (z^*)^l U(C)$. Since z is a center element, the non-abelian magnetic "flux" created by such singular gauge transformations is quantized, and the magnetic objects themselves are known as "thin center vortices." As in QED, the singular magnetic field on the $D - 2$ dimensional hypersurface can be smeared out in the directions transverse to the hypersurface into a region of finite thickness, much like the field of a solenoid. This is a *thick* center vortex.

The center vortex creation operator is closely related to the center symmetry transformation (4.12). In $D = 2$ dimensions we can create a thin vortex at the plaquette at site x_0, y_0 by the following operation (Fig. 4.2):

$$U_y(x, y_0) \rightarrow zU_y(x, y_0) \quad \text{for } x > x_0 . \tag{4.36}$$

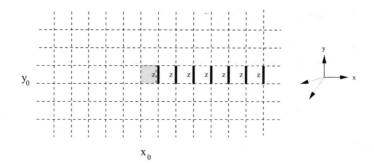

Fig. 4.2 Creation of a thin center vortex. The shaded plaquette, and all other x-y plaquettes at sites $(x_0, y_0, \mathbf{x}_\perp)$ form the center vortex. The stack of vortex plaquettes lie along a line in $D = 3$ dimensions, or a surface in $D = 4$ dimensions

In higher dimensions, the operation

$$U_y(x, y_0, x_\perp) \to z U_y(x, y_0, x_\perp) \quad \text{for } x > x_0 \tag{4.37}$$

creates a line of magnetic flux parallel to the z-axis ($D = 3$), or a surface of magnetic flux parallel to the $z - t$ plane ($D = 4$).

In the Hamiltonian formulation of $3 + 1$ dimensional gauge theory, we may consider an operator $B(C)$ which creates a thin center vortex along loop C at a particular time t. Then if C and C' are loops at a given time t, and if the two loops are linked (with unit linking number) in the $D = 3$ dimensional timeslice, we have

$$B(C)U(C') = zU(C')B(C) \quad , \quad z \in Z_N . \tag{4.38}$$

Using only this relation, 't Hooft [4] argued that

- only an area-law or a perimeter-law falloff for $W[C]$ and $\langle B[C] \rangle$ is possible; and
- in the absence of massless excitations, it is impossible to have both

$$W(C) \sim e^{-aP(C)} \quad \text{and} \quad \langle B(C) \rangle \sim e^{-bP(C)} . \tag{4.39}$$

So a perimeter law falloff for $B(C)$ implies an area-law falloff for $W(C)$, i.e. confinement. $B(C)$ is known as the '*t Hooft loop operator*. It is a creation operator for thin center vortices, which (at fixed time t) are loops of color magnetic flux, and can be thought of as being in some sense "dual" to the Wilson loop operator, which (at fixed time) creates a closed loop of color electric flux.

4.4   The Vortex Free Energy

't Hooft devised another procedure for inserting a center vortex into a finite lattice
[5]. Consider $SU(N)$ gauge theory on a finite two-dimensional lattice. Pick one
plaquette p', with

$$U(p') = U_1(x_0, y_0)U_2(x_0 + 1, y_0)U_1^\dagger(x_0, y_0 + 1)U_2^\dagger(x_0, y_0) , \qquad (4.40)$$

and modify the action by replacing $U(p')$ by $zU(p')$, where $z \in Z_N$, i.e.

$$S = -\frac{\beta}{2N}\left\{\sum_{p \neq p'}(\text{Tr}[U(p)] + c.c.) + (\text{Tr}[zU(p')] + c.c.)\right\} . \qquad (4.41)$$

On a finite lattice this replacement can be formulated as a modification of the ordinary periodic boundary conditions, and is known as "twisted boundary conditions."
Twisted boundary conditions can also be regarded as a shift $\beta \to z\beta$ in the coupling
constant at the plaquette p'. It is clear that setting all links equal to the identity
matrix (or a gauge transformation of the identity matrix) no longer minimizes the
action. On an infinite lattice, the action would be minimized by any configuration
gauge-equivalent to

$$U_2(x, y_0) = z\mathbb{1}_N \text{ for } x \leq x_0 , \qquad (4.42)$$

with all other links set equal to the $N \times N$ identity matrix $\mathbb{1}_N$. This is a Dirac line,
running parallel to the x-axis from $x = x_0$ to $x = -\infty$. On a finite lattice the line
has to end somewhere; it must end on a center vortex. There is no need, however,
for the line to end on a thin center vortex, with all of the center flux contained in a
single plaquette. That arrangement costs a great deal of action at large β. The action
is lowered if the center flux is spread over many plaquettes, in a region of some
finite extension.

In $D = 4$ dimensions, the twisted boundary conditions involve replacing $U(p')$
by $zU(p')$ in the action, for all plaquettes p' oriented in the $x - y$ plane, at fixed
x_0, y_0 and all z, t, i.e.

$$U(p') = U_1(x_0, y_0, z, t)U_2(x_0 + 1, y_0, z, t)U_1^\dagger(x_0, y_0 + 1, z, t)U_2^\dagger(x_0, y_0, z, t) . \qquad (4.43)$$

This replacement then creates a thick center vortex sheet somewhere on the lattice,
oriented parallel to the $z - t$ plane and closed by ordinary periodic boundary
conditions in the z and t directions. The actual thickness of the vortex is not
determined by the $U(p') \to zU(p')$ replacement; it is instead a dynamical issue. Of
course, there is an upper limit to the thickness of a center vortex on a finite lattice,
the limit is the cross-sectional area of the lattice in the $x - y$ plane.

For simplicity, consider the SU(2) gauge group, with $z = -1$ for twisted boundary conditions. We define Z_+ as the lattice partition function with ordinary boundary conditions, and Z_- as the partition function with twisted boundary conditions. The magnetic free energy F_{mg} of a Z_2 vortex is defined by

$$e^{-F_{mg}} = \frac{Z_-}{Z_+} \, . \tag{4.44}$$

The electric free energy F_{el} is defined by a Z_2 Fourier transform

$$e^{-F_{el}} = \sum_{z=\pm} z \frac{Z_z}{Z_+}$$

$$= 1 - e^{-F_{mg}} \, . \tag{4.45}$$

Let C be a rectangular loop of area $A[C]$. The following inequality was proven by Tomboulis and Yaffe in [6]

$$W(C) \le \{\exp[-F_{el}]\}^{A(C)/(L_x L_y)} \, . \tag{4.46}$$

So if the vortex magnetic free energy falls off with cross-sectional area $L_x L_y$ as

$$F_{mg} \sim L_z L_t e^{-\rho L_x L_y} \, , \tag{4.47}$$

then for large L_x, L_y we have $F_{mg} \ll 1$ and $\exp[-F_{el}] \approx F_{mg}$. The behavior (4.47) is therefore a sufficient condition for confinement, because it implies, from (4.46), an area law bound for Wilson loops.

Numerical investigations of these quantities were begun by Kovács and Tomboulis [7]; their computation of Z_-/Z_+ is shown in Fig. 4.3, and is consistent with the vortex magnetic free energy falling very rapidly to zero on large lattice volumes, in accordance with the condition (4.47). Much additional work along these lines was carried out by de Forcrand and von Smekal [8–10]. Note that, in Fig. 4.3, the ratio Z_-/Z_+ attains its limiting value of unity for lattice extensions of roughly 1.25 fm. It appears that the free energy of a vortex is lowered by allowing the vortex to spread out to this extent, but there is no advantage in spreading to a thickness much beyond that. This suggests that the thickness of a center vortex on a large lattice is about 1.25 fm.

4.5 Summary

To summarize the lessons of this chapter: We have seen that the existence of a non-vanishing asymptotic string tension requires that the gauge group has a non-trivial center, and that any matter fields must transform trivially with respect to the center subgroup. When these two conditions are satisfied, the action of the gauge theory

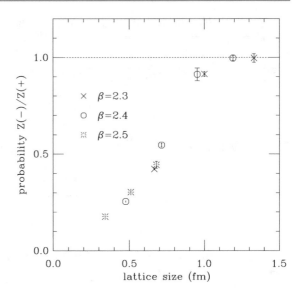

Fig. 4.3 The behavior of the vortex free energy vs. lattice extension (same in all directions). From Kovács and Tomboulis [7]

is invariant with respect to a global center symmetry transformation, which is not a gauge transformation. All of the signals we have seen for confinement

- an area law for Wilson loops;
- vanishing Polyakov lines;
- perimeter-law 't Hooft loops;
- area-law falloff for the vortex magnetic free energy;

can only be satisfied if a non-trivial global center symmetry exists, and is not spontaneously broken.

This motivates the idea that the vacuum fluctuations responsible for the asymptotic string tension must be associated, in some way, with center symmetry.

References

1. K.G. Wilson, Confinement of quarks. Phys. Rev. D **10**, 2445 (1974)
2. R. Narayanan, H. Neuberger, Infinite N phase transitions in continuum Wilson loop operators. J. High Energy Phys. **0603**, 064 (2006). arXiv:hep-th/0601210
3. A.M. Polyakov, Compact gauge fields and the infrared catastrophe. Phys. Lett. B **59**, 82 (1975)
4. G. 't Hooft, On the phase transition towards permanent quark confinement. Nucl. Phys. B **138**, 1 (1978)
5. G. 't Hooft, A property of electric and magnetic flux in non-Abelian gauge theories. Nucl. Phys. B **153**, 141 (1979)
6. E. Tomboulis, L. Yaffe, Finite temperature SU(2) lattice gauge theory. Commun. Math. Phys. **100**, 313 (1985)
7. T. Kovács, E. Tomboulis, Computation of the vortex free energy in SU(2) gauge theory. Phys. Rev. Lett. **85**, 704 (2000). arXiv:hep-lat/0002004

8. L. von Smekal, Ph. de Forcrand, 't Hooft loops, electric flux sectors and confinement in SU(2) Yang-Mills theory. Phys. Rev. D **66**, 011504 (2002). arXiv:hep-lat/0107018
9. L. von Smekal, Ph. de Forcrand, Electric and magnetic fluxes in SU(2) Yang-Mills theory. Nucl. Phys. Proc. Suppl. **119**, 655 (2005). arXiv:hep-lat/0209149
10. L. von Smekal, Ph. de Forcrand, O. Jahn, More on electric and magnetic fluxes in SU(2) (2002). arXiv:hep-lat/0212019

Properties of the Confining Force

5

Abstract

The properties of the confining force: asymptotic linearity, Casimir scaling at intermediate scales, N-ality dependence asymptotically, and the Lüscher term.

Any fully satisfactory theory of confinement ought to explain the following features of the confining force, for which there is strong theoretical and/or numerical support:

1. asymptotic linearity of the static potential;
2. Casimir scaling of string tensions at intermediate distance scales;
3. N-ality dependence of asymptotic string tensions;
4. evidence of quantum string-like behavior: roughening and the Lüscher term.

Let's begin with asymptotic linearity. It was proven by Bachas in [1] that the static quark potential must be everywhere convex, i.e.

$$\frac{dV}{dR} > 0 \quad \text{and} \quad \frac{d^2V}{dR^2} \leq 0 . \tag{5.1}$$

The inequality is saturated by a linear potential; $V(R)$ can be bounded from above by a straight line. Monte Carlo simulations provide very convincing evidence that the static quark potential is indeed linear asymptotically. A sample numerical calculation of $V(R)$ in SU(3) gauge theory, in units of the "Sommer scale" $r_0 \approx 0.5$ fm, is shown in Fig. 5.1.

Casimir scaling is the property that at some intermediate range of distances, the slope of the static quark potential for a quark in group representation r, and its antiquark, is approximately proportional to the quadratic Casimir C_r of the

© Springer Nature Switzerland AG 2020
J. Greensite, *An Introduction to the Confinement Problem*, Lecture Notes
in Physics 972, https://doi.org/10.1007/978-3-030-51563-8_5

Fig. 5.1 The static quark potential in SU(3) lattice gauge theory, normalized to $V(r_0) = 0$. The reference distance r_0 is the "Sommer scale," which roughly 0.5 fm. From Bali [2]

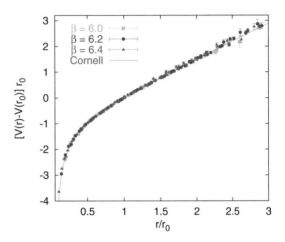

representation; i.e.

$$\sigma_r = \frac{C_r}{C_F}\sigma_F \, , \tag{5.2}$$

where F denotes the fundamental, defining representation. This relationship is certainly true at weak couplings in $D = 2$ dimensional gauge theory, where Wilson loops can be calculated from one-gluon exchange. The proposal that Casimir scaling might also hold in higher dimensions [3] is based on the notion of "dimensional reduction", which suggests that the calculation of a large planar Wilson loop in $D = 4$ dimensions reduces somehow to the corresponding calculation in $D = 2$ dimensions. At strong couplings, dimensional reduction is demonstrably correct; the leading term for the string tension derived from the strong-coupling expansion is the two-dimensional result. On the other hand, the non-leading terms *do* depend on dimension. Arguments for why dimensional reduction might hold nonetheless, at weak couplings, were advanced independently by Olesen [4] and myself [5], and the interested reader can consult those references. In any case, Casimir scaling in SU(2) and SU(3) gauge theory seems to hold with remarkable precision. Figure 5.2 shows the data for string tensions of the static quark potential, in $SU(3)$ lattice gauge theory, in various representations of the color group. The solid lines shown for each representation is obtained by multiplying a fit to the static potential of the fundamental (3) representation by the ratio of quadratic Casimirs C_r/C_3.[1] It can be argued that Casimir scaling must hold *exactly* for the $SU(\infty)$ gauge group; we will reserve that argument for Chap. 11, which is concerned with the large-N limit.

---

[1] It should be noted that the calculation on which this figure is based uses a method which creates metastable flux tubes, which are then allowed to propagate for a relatively short Euclidean time interval. This procedure is insensitive to the string-breaking process, and hence one can only calculate the string tension of the metastable states.

Fig. 5.2 Numerical evidence
for Casimir scaling in SU(3)
lattice gauge theory. The solid
lines are obtained from a fit of
the potential in fundamental
representation, multiplied by
a ratio of quadratic Casimirs
C_r/C_F. The reference
distance r_0 is again about 0.5
fm. From Bali [6]

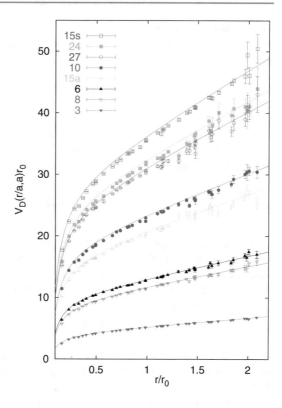

Casimir scaling cannot continue out to arbitrary quark-antiquark separations; apart from $D = 2$ dimensions or $N = \infty$ colors it is an intermediate distance phenomenon. Consider a quark-antiquark pair in representation r, with N-ality k_r. Gluons can bind to the quark and antiquark, and reduce the color charge to the lowest-dimensional representation with the same N-ality. It follows that after screening by gluons (i.e. asymptotically), string tensions depend only on the N-ality of the representation

$$\sigma_r = \sigma(k_r) . \tag{5.3}$$

String tensions of the lowest-dimensional representation of N-ality k are often called "k-string tensions."

It is not known precisely how k-string tensions depend on k; two proposals are so-called "Casimir scaling", and Sine-law scaling. "Casimir scaling", in this context, is something of a misnomer. The term as originally introduced by Del Debbio, Faber, Olejnik and myself [7, 8] refers to intermediate string tensions (prior to screening by gluons), and is defined in (5.2) above. Somewhat confusingly, Casimir scaling has also been used in the literature in connection with k-string tensions, which refer to string tensions at asymptotic, rather than intermediate, distances. In that usage, the Casimir C_r is replaced by the Casimir of the lowest dimensional representation with the same N-ality as r. Then, for an SU(N) gauge theory, the "Casimir scaling" prediction is

$$\sigma_r = \sigma(k_r) = \frac{k(N-k)}{N-1}\sigma_F \,, \tag{5.4}$$

where σ_F is the string tension for quarks in the fundamental, defining representation.

An alternative proposal, motivated by supersymmetric gauge theories, is the Sine-law dependence

$$\sigma_r = \sigma(k_r) = \frac{\sin(\pi k/N)}{\sin(\pi/N)}\sigma_F \,. \tag{5.5}$$

In practice, the Casimir scaling and Sine-law values for the k-string tensions are not very different, and there has been some debate in the literature over whether the Sine law is favored. For our purposes, the crucial point is not Casimir scaling vs. the Sine Law, but rather the simple fact that asymptotic string tension depends only on the N-ality of the representation.

An important example is the asymptotic string tension of the adjoint representation, which has $k = 0$. Gluons can screen adjoint color charges and the confining string should break, leaving two bound states known as "gluelumps", which can be thought of as a gluon bound to a static quark in the color adjoint representation. Suppose the gluonic energy of a gluelump state is m_{GL}. Then the adjoint string should break for quark separations R such that

$$2m_{GL} < \sigma_A R \,, \tag{5.6}$$

where σ_A is the adjoint string tension, which is approximately equal to $C_A \sigma_F / C_F$ in the intermediate region, according to Casimir scaling. Beyond the string-breaking distance the static quark potential is flat, and equal to $2m_{GL}$.

To observe string-breaking (i.e. color screening by gluons) for adjoint representation heavy quark states, using only rectangular Wilson $R \times T$ Wilson loops, is computationally demanding, but was achieved in numerical simulations by de Forcrand and Kratochvila [9], using a noise reduction technique due to Lüscher and Weisz [10]. The result of their calculation, for SU(2) lattice gauge theory in $D = 3$ dimensions, is shown in Fig. 5.3. The adjoint string breaks at about ten lattice spacings, corresponding (at the given value of the lattice coupling) to 1.25 fm in physical units. We can see that Casimir scaling of the adjoint potential is fairly good up to 10 lattice spacings, at $\beta = 6.0$, after which the potential abruptly flattens out.

Although I have used the terms "QCD string" and "string breaking", there is the question of whether, or to what extent, the QCD flux tube really behaves like a quantum string as described by string theory. If the QCD flux tube resembles a Nambu string (see Chap. 14), then transverse fluctuations of the string induce a

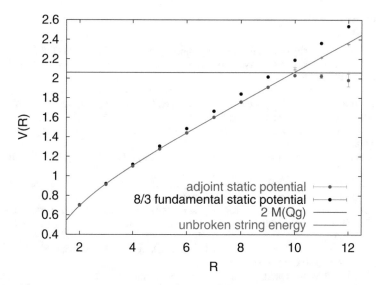

Fig. 5.3 The adjoint and $\frac{8}{3}\times$fundamental static potentials $V(R)$ vs R, in $D = 3$ dimensional SU(2) lattice gauge at $\beta = 6.0$. The horizontal line at 2.06(1) represents twice the energy of a gluelump. From de Forcrand and Kratochvila [9]

universal (coupling and scale independent) $1/R$ modification to the linear potential [11, 12] in D-dimensions

$$V(R) = \sigma R - \frac{\pi(D-2)}{24}\frac{1}{R} + c .\qquad (5.7)$$

This modification is known to lattice gauge theorists as the "Lüscher term."[2] Another prediction of the Nambu model is that the cross-section of the QCD flux tube is not actually constant, but should grow logarithmically with quark separation, a phenomenon known as "roughening" [13, 14]. There is some evidence in favor of both the Lüscher term [10, 15] and logarithmic growth of the flux tube cross-section [16], as well as a spectrum of flux tube excited states of which also appears to correspond to that of a string theory [17]. Perhaps the most accurate calculations to date [18] have been carried out in $2 + 1$ dimensions, for the spectrum of closed flux tubes of length R which wind once around the periodic lattice in a spatial direction. Both the R-dependence of the energy of the ground state (which includes the Lüscher term), and the pattern of excited flux tube states, seem to agree rather well with predictions of the Nambu string action.

---

[2]To add to the "Casimir" confusion, the Lüscher term is known to string theorists as the "Casimir energy."

References

1. C. Bachas, Convexity of the quarkonium potential. Phys. Rev. D **33**, 2723 (1986)
2. G. Bali, QCD forces and heavy quark bound states. Phys. Rept. **343**, 1 (2001). arXiv: hep-ph/0001312
3. J. Ambjorn, P. Olesen, C. Peterson, Stochastic confinement and dimensional reduction: (1). Four-Dimensional SU(2) lattice gauge theory. Nucl. Phys. B **240**, 189 and 533 (1984)
4. P. Olesen, Confinement and random fluxes. Nucl. Phys. B **200** [FS4], 381 (1982)
5. J. Greensite, Calculation of the Yang-Mills vacuum wave functional. Nucl. Phys. B **158**, 469 (1979). Nucl. Phys. B166, 113 (1980)
6. G. Bali, Casimir scaling of SU(3) static potentials. Phys. Rev. D **62**, 114503 (2000). arXiv: hep-lat/0006022
7. L. Del Debbio, M. Faber, J. Greensite, Š. Olejník, Casimir scaling versus Abelian dominance in QCD string formation. Phys. Rev. D **53**, 5891 (1996). arXiv: hep-lat/9510028
8. L. Del Debbio, M. Faber, J. Greensite, Š. Olejník, Some cautionary remarks on Abelian projection and Abelian dominance. Nucl. Phys. Proc. Suppl. **53**, 141 (1997). arXiv: hep-lat/9607053
9. P. de Forcrand, S. Kratochvila, Observing string breaking with Wilson loops. Nucl. Phys. Proc. Suppl. **119**, 670. arXiv: hep-lat/0209094
10. M. Lüscher, P. Weisz, Quark confinement and the bosonic string. JHEP **07**, 049 (2002). arXiv: hep-lat/0207003
11. M. Lüscher, Symmetry breaking aspects of the roughening transition in gauge theories. Nucl. Phys. B **180**(FS2), 317 (1981)
12. O. Alvarez, The static potential in string models. Phys. Rev. D **24**, 440 (1981)
13. M. Lüscher, G. Münster, P. Weisz, How thick are chromoelectric flux tubes? Nucl. Phys. B **180**(FS2), 1 (1981)
14. A. Hasenfratz, E. Hasenfratz and P. Hasenfratz, Generalized roughening transition and its effect on the string tension. Nucl. Phys. B **180**, 353 (1981)
15. K.J. Juge, J. Kuti, C. Morningstar, QCD string formation and the Casimir energy. arXiv: hep-lat/0401032
16. V.G. Bornyakov, A.V. Kovalenko, M.I. Polikarpov, D.A. Sigaev, Confining string and P-vortices in the indirect Z(2) projection of SU(2) lattice gauge theory. Nucl. Phys. Proc. Suppl. **119**, 739 (2003). arXiv:hep-lat/0209029
17. J. Kuti, QCD and string theory. PoS **LAT2005**, 001 (2006). arXiv:hep-lat/0511023
18. A. Athenodorou, B. Bringoltz, M. Teper, The closed string spectrum of SU(N) gauge theories in $2 + 1$ dimensions. Phys. Lett. B **656**, 132 (2007). arXiv:0709.0693 [hep-lat]

Confinement from Center Vortices I

6

Abstract

The center vortex confinement mechanism. Methods for locating center vortices in lattice configurations. Numerical evidence that center vortices are responsible for the asymptotic string tension.

We have already seen a number of indications that the center of the gauge group has something to do with confinement. For one thing, the existence of a non-vanishing asymptotic string tension is tied to the existence, and unbroken realization, of a global center symmetry. Moreover, two of the order parameters for confinement, namely the 't Hooft loop $B(C)$ and vortex free energy, are explicitly associated with center vortex creation. But perhaps the strongest motivation for the center vortex mechanism of confinement, to be discussed below, comes from the fact that the asymptotic string tension depends only on the N-ality of the quark charge. Whatever vacuum fluctuations of the gauge field are responsible for an asymptotic string tension for quarks in the fundamental representation, those same fluctuations should not also give rise to an asymptotic string tension for quarks in the adjoint representation, or in any other zero N-ality representation of the gauge group.

6.1 The Mechanism

There is, of course, no great mystery in the fact that the adjoint string tension vanishes asymptotically. We have already seen why that must be true. As the flux tube gets longer, at some stage it is energetically favorable to pair-create some gluons which bind to each adjoint quark to form a singlet, and the flux tube breaks. This explanation is certainly correct, but when we talk of pair-creating gluons out the vacuum, we are using the language of Feynman diagrams, and talking about particle states. However, the path-integral itself is a sum over field configurations.

© Springer Nature Switzerland AG 2020

J. Greensite, *An Introduction to the Confinement Problem*, Lecture Notes in Physics 972, https://doi.org/10.1007/978-3-030-51563-8_6

When a Wilson loop is evaluated at a given location by Monte Carlo methods, the configurations which are generated have no knowledge whatever of the location of the Wilson loop being measured, or where the pair-created gluons should appear. The Wilson loop is just an observable, not a part of the action. Nevertheless, the Monte Carlo evaluation must somehow give us the same answer that we deduce, on the basis of energetics, from the particle picture of string breaking. So this raises the question: What feature of a typical vacuum field fluctuation can give us both a finite string tension for quarks in the fundamental representation, and a vanishing string tension for zero N-ality quark charges, at large distances? What we are looking for is a "field" explanation of screening, to complement the particle picture.

Suppose that there is a certain type of solitonic object, identifiable in typical field configurations, whose fluctuations in position (or some other degree of freedom) give rise to an area law for Wilson loops. In order that the string tension depend only on N-ality, it seems necessary that Wilson loops which are in different representations, but have the same N-ality, should be affected by these solitonic objects in exactly the same way. Loops of zero N-ality should not be affected at all. These requirements are almost a definition of center vortices; I do not know of any other object with the same properties. Suppose we create a center vortex linked to loop C. This affects the loop holonomy as follows: $U(C) \rightarrow zU(C)$ where $z \in Z_N$ for an $SU(N)$ gauge group. Then, for a Wilson loop in representation r with N-ality k

$$W_r(C) \rightarrow z^k W_r(C) , \tag{6.1}$$

so that Wilson loops of the same N-ality are, indeed, affected in the same way, while zero N-ality loops are unchanged. Creation of a set of vortices, with center elements $z_a, z_b, z_c \ldots$, linked to loop C, with linking numbers $l_a, l_b, l_c \ldots$ respectively, will transform the loop as

$$W_r(C) \rightarrow Z^k(C)W_r(C)$$
$$Z(C) = z_a^{l_a} z_b^{l_b} z_c^{l_c} \ldots . \tag{6.2}$$

The center vortex scenario for confinement [1–5] is very simple: Any vacuum (thermalized lattice) configuration can be viewed as a set of center vortices superimposed on a non-confining configuration, and the area law for Wilson loops is obtained from random fluctuations in the number of vortices (and their linking numbers) topologically linked to the loop. Let the gauge group be SU(2) for simplicity. Consider a plane of area L^2 (Fig. 6.1) which is pierced, at random locations, by N center vortices, and consider a Wilson loop of area A in any half-integer group representation lying in that plane. Then the probability that n of those N vortices will lie inside the area A is

$$P_N(n) = \binom{N}{n} \left(\frac{A}{L^2}\right)^n \left(1 - \frac{A}{L^2}\right)^{N-n} \tag{6.3}$$

Fig. 6.1 A plane of area L^2, pierced by N center vortices (dots). A Wilson loop of area A lies in the plane

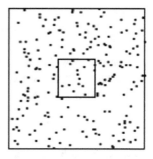

Each vortex piercing the Wilson loop contributes a factor of -1, so the vortex contribution to the Wilson loop is

$$W(C) = \sum_{n=0}^{N} (-1)^n P_N(n) = \left(1 - \frac{2A}{L^2}\right)^N \tag{6.4}$$

Now keeping the vortex density $\rho = N/L^2$ fixed, and taking the $N, L \to \infty$ limit, we arrive at the Wilson loop area law falloff

$$W(C) = \lim_{N \to \infty} \left(1 - \frac{2\rho A}{N}\right)^N = e^{-2\rho A} \tag{6.5}$$

That is the vortex confinement mechanism in three lines [6]. I believe it is the simplest known. The crucial assumption is that vortex piercings in the plane are random and uncorrelated, and this implies that vortices percolate throughout the spacetime volume.

An alternate derivation on the lattice, for any gauge group, goes as follows: We begin by writing the holonomy in the form

$$U(C) = Z(C)u(C) , \tag{6.6}$$

where $u(C)$ is the contribution to the holonomy from the non-confining background. Then the idea is that

$$\langle \chi_r[U(C)] \rangle \approx \langle Z^k(C) \rangle \langle \chi_r[u(C)] \rangle$$
$$\approx e^{-\sigma_k A(C)} e^{-\mu_r P(C)} . \tag{6.7}$$

To get this result, we only need that

A) For a single large loop C, $Z(C)$ and $u(C)$ are weakly correlated; and
B) for two large loops C_1 and C_2, $Z(C_1)$ and $Z(C_2)$ are weakly correlated; i.e.

$$\langle Z(C_1)Z(C_2) \rangle \approx \langle Z(C_1) \rangle \langle Z(C_2) \rangle \tag{6.8}$$

and in fact this weak correlation applies to the product of any number of large loops lying in the same plane.

To see how we arrive at an area law from these assumptions, consider a large rectangular Wilson loop C of area A, in group representation r of N-ality k. We have, by assumption A,

$$W_r(C) = \langle \chi_r[U(C)] \rangle = \langle Z^k(C) \rangle \langle \chi_r[u(C)] \rangle . \tag{6.9}$$

Now subdivide the area A into square $L \times L$ subareas bounded by loops $\{C_i\}$, so that

$$\langle Z^k(C)] \rangle = \langle \prod_{i=1}^{A/L^2} Z^k(C_i) \rangle$$

$$\approx \prod_{i=1}^{A/L^2} \langle Z^k(C_i) \rangle$$

$$= \exp[-\sigma(k)A(C)] , \tag{6.10}$$

where assumption B is used in going from the first to second line, and

$$\sigma(k) = -\frac{\log\left[\langle Z^k(C_i) \rangle\right]}{L^2} \tag{6.11}$$

for any of the $L \times L$ subloops C_i. Solving for $\langle Z^k(C_i) \rangle$, we have

$$\langle Z^k(C_i) \rangle = \exp[-\sigma(k)L^2] . \tag{6.12}$$

which means that the string tension extracted from the smaller loops is identical to that of the larger loop, providing the smaller loops are large enough so that assumption B holds, i.e. providing the larger loop can be subdivided into smaller loops such that the $Z(C_i)$ fluctuate independently. Note that we have simply applied the reasoning given in Sect. 2.4 to the center-valued loops $Z(C)$.

It is interesting to consider the limiting case, where $Z(p)$ fluctuates independently from one plaquette p on the lattice to another, for plaquettes lying in the same plane. Start from

$$Z(C) = \prod_{p \in A(C)} Z(p) , \tag{6.13}$$

where $A(C)$ is the minimal area of the loop. For simplicity consider the group $SU(2)$, where

$$Z(p) = \begin{cases} -1 \text{ if the vortex pierces the plane at } p \\ +1 \text{ otherwise} \end{cases}, \tag{6.14}$$

and define f to be the probability that $Z(p) = -1$ on any given plaquette p. Then, assuming the $Z(p)$ fluctuate independently,

$$\langle Z(C) \rangle = \langle \prod_{p \in A(C)} Z(p) \rangle = \prod_{p \in A(C)} \langle Z(p) \rangle$$

$$= \prod_{p \in A(C)} \left[(1 - f)(+1) + f(-1) \right]$$

$$= \exp\left[A(C) \ln(1 - 2f) \right], \tag{6.15}$$

and the string tension is

$$\sigma = -\ln(1 - 2f). \tag{6.16}$$

Now, what kind of link configurations produce holonomies, for any C, which are Z_N elements $U(C) = Z(C) \in Z_N$? The answer is that these are configurations which can be transformed, by an $SU(N)$ gauge transformation, into link configurations of Z_N lattice gauge theory, i.e.

$$\mathcal{U}_\mu(x) = z_\mu(x) g(x) g^{-1}(x + \widehat{\mu}) \quad, \quad z_\mu(x) \in Z_N. \tag{6.17}$$

The excitations of Z_N lattice gauge theory are precisely thin Z_N vortices. Then we may subdivide link variables into degrees of freedom responsible for confinement $z_\mu(x)$, and those which are simply a non-confining background $V_\mu(x)$, as follows[1]

$$U_\mu(x) = g(x) z_\mu(x) V_\mu(x) g^{-1}(x + \widehat{\mu}). \tag{6.18}$$

This form motivates us to look for the gauge transformation $g(x)$ which would bring a lattice configuration $\mathcal{U}_\mu(x)$ into the form of a simple product $U = zV$. If V, the non-confining background, is (mainly) a small fluctuation around the unit element then we can easily determine the large fluctuations $z_\mu(x)$, and, from those, the center vortex locations.

---

[1] $V_\mu(x)$ would also be responsible for "thickening" the flux of the thin vortices, so there must be some short range correlation between $V_\mu(x)$ and vortex position.

6.2   Center Gauges and Center Projection

The vortex mechanism is probably the simplest route to confinement, and is well motivated by the local, gauge-invariant order parameters for confinement ('t Hooft loop, Polyakov loop, vortex free energy), and by the known facts about N-ality dependence. But is it right?

To find out, we turn to lattice Monte Carlo simulations. The first problem is to figure out how to spot thick center vortices in a list of what looks like random numbers, i.e. the lattice link variables.

6.2.1   Direct Maximal Center Gauge

We would like to transform the link variables to a gauge in which link variables can be factored into center elements, whose fluctuations bring about the asymptotic string tension, and non-confining fluctuations $V_\mu(x)$, which have no influence on the far-infrared physics. But what gauge is that? The most intuitive choice [7] is to transform the configuration so that the link variables $U_\mu(x)$ are as close as possible to some Z_N lattice configuration or, to put it another way, the link variables in the adjoint representation are as close as possible to unity. The center elements of $SU(N)$ are all mapped to the identity in the adjoint representation, so the deviation of adjoint representation links from the identity matrix is a measure of the deviation of the lattice configuration from a pure Z_N configuration. The gauge in which this deviation is minimized is known as *"direct maximal center gauge."* It is defined as the gauge which maximizes

$$R = \sum_x \sum_\mu \text{Tr}[U_{A\mu}(x)]$$

$$= \sum_x \sum_\mu \text{Tr}[U_\mu(x)]\text{Tr}[U_\mu^\dagger(x)] - 1 \,, \tag{6.19}$$

where $U_{A\mu}$ is the link variable in the adjoint representation. This is simply Landau gauge in the adjoint representation. In this gauge we decompose

$$U_\mu(x) = Z_\mu(x)V_\mu(x) \,, \tag{6.20}$$

where $Z_\mu(x)\mathbb{1}_N$ is the center element in the $SU(N)$ group manifold which is closest to $U_\mu(x)$. For example, in $SU(2)$,

$$Z_\mu(x) = \text{signTr}[U_\mu(x)] \,. \tag{6.21}$$

In general, in an $SU(N)$ gauge theory, we take $Z_\mu(x) = z_n$, where $z_n \in Z_n$ is the element for which $\text{ReTr}[z_n^* U_\mu(x)]$ is largest.

The mapping of the gauge-fixed $SU(N)$ lattice configuration to a Z_N lattice configuration

$$U_\mu(x) \to Z_\mu(x) \tag{6.22}$$

is known as *"center projection."* Plaquettes with $Z(p) \neq 1$ on the projected lattice, where $Z(p)$ is the product of $Z_\mu(x)$ link variables around plaquette p, are known as *"P-plaquettes,"* and together they identify the position of thin center vortices known as *"P-vortices."* The claim is that this procedure locates center vortices on the unprojected lattice; P-vortices lie somewhere in the middle of the thick vortices of the original lattice configuration.

P-vortices, strictly speaking, are not located on the original lattice, but rather on the corresponding "dual" lattice, whose sites are shifted away from the sites of the original lattice by half a lattice spacing in the $\mu = 1, \ldots, D$ directions. In $D = 2$ dimensions, a plaquette is said to "dual to" (intersected in the middle by) a site of the dual lattice, located at the center of the plaquette. In three dimensions, a plaquette is dual to a link on the dual lattice, orthogonal to the plaquette, which runs through the center of the plaquette. In four dimensions a plaquette is dual to a plaquette on the dual lattice, which is oriented in a plane orthogonal to original plaquette; the areas of the two plaquettes intersect at a common midpoint. Suppose, for example, that at a fixed time we have a set of P-plaquettes oriented orthogonal to the x-axis, as shown in Fig. 6.2. A P-vortex in this timeslice is a line which pierces the center of each P-plaquette, and carries magnetic flux in the center of the gauge group. As the vortex line propagates in time, it traces out an area on the dual lattice, formed by plaquettes which are dual to the P-plaquettes.

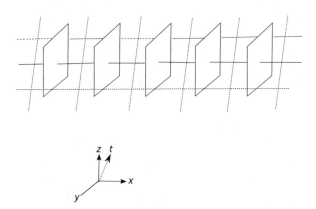

Fig. 6.2 A set of P-plaquettes, oriented parallel to the y-z plane. A thin vortex in $D = 3$ dimensions is a line running through the middle of the P-plaquettes, carrying a unit of center flux. The vortex line is dual to the P-plaquettes and runs along links of the dual lattice. In $D = 4$ dimensions the thin vortex is a surface on the dual lattice, indicated the dashed lines, formed by plaquettes which are dual to the P-plaquettes. Note that the sites of the dual lattice are displaced by half a lattice spacing in time, as well as in the space directions

The center-projected lattice is a configuration of Z_N lattice gauge theory, which of course is an abelian gauge theory, and has a simple Stokes Law

$$Z(C) = \prod_{p \in S(C)} Z(p) , \qquad (6.23)$$

where $S(C)$ is any surface bounded by C, and the product is over the plaquettes p which make up that surface. Let C be a planar loop, and $S(C)$ be the minimal surface. If there are no P-plaquettes in the minimal surface, then $Z(C) = 1$. If there is one or more, then the product can be restricted to the P-plaquettes ($Z(p) \neq 1$) only,

$$Z(C) = \prod_{\substack{p \in S(C) \\ Z(p) \neq 1}} Z(p) . \qquad (6.24)$$

Since each P-plaquette is dual to a plaquette (four dimensions) or a link (three dimensions) on a P-vortex which intersects the P-plaquette at a point in the middle, we say that the minimal surface is "pierced" by a P-vortex at the location of the P-plaquette.

Let us refer to the maximum distance between any two points on a vortex as the *extension* of the vortex, and suppose that there were an upper limit L_{max} to vortex extension. For very large Wilson loops, this means that the only vortices which can link to the loop are those which are in the neighborhood (i.e. within a distance L_{max}) of the loop perimeter, and the minimal area is irrelevant. It is not too hard to see that a perimeter, rather than an area-law falloff of center-projected Wilson loops $\langle Z(C) \rangle$ would be the result (cf. [8]). This means that if the center-projected loops do, in fact, have an area-law falloff, then the only possible upper bound on vortex extension is the finite lattice volume. The picture is that P-vortices are random surfaces which "percolate" throughout the spacetime volume. The piercings, by such vortices, of widely separated plaquettes on any given planar surface are uncorrelated, leading to a lack of correlation among large center-projected loops.

6.2.2 Finding Thin Vortices

If the gauge-fixing and center projection procedure really locates center vortices on the unprojected lattice, whose position is initially unknown, then it should also locate vortices whose position is known beforehand. In other words, suppose we have some lattice configuration U, and insert by hand a center vortex, via a singular gauge transformation, somewhere in the lattice. Assuming it is possible to gauge transform to a unique global maximum of R, will this center vortex be among the set of vortices identified by the center projection procedure? The answer is yes, and the argument is as follows: Let U denote some thermalized lattice configuration. A thin center vortex is created, on the background U, by a singular gauge transformation. Denote the resulting configuration as U'. Since the discontinuity of the singular gauge transformation is a center element, this discontinuity is invisible in the adjoint representation. It follows that the two lattice configurations in the adjoint representation, U_A and U'_A, are gauge equivalent. Then maximal center gauge will take U_A and U'_A into the *same* gauge-fixed configuration $\widetilde{U}_A$. In the passage to maximal center gauge, the link variables U and U' in the fundamental representation are transformed to configurations $\widetilde{U}$ and $\widetilde{U}'$ which both equal $\widetilde{U}_A$ in the adjoint representation. Since they correspond to the same configuration in the adjoint representation, the $\widetilde{U}$ and $\widetilde{U}'$ lattice configurations can differ only by continuous and/or singular Z_N transformations. This means that the singular $SU(N)$ gauge transformation which inserted the thin vortex in U' must appear as a singular Z_N transformation relating $\widetilde{U}'$ to $\widetilde{U}$.

What has happened here is that the original singular $SU(N)$ gauge transformation, which may be quite smooth (up to the discontinuity) and extended, has been squeezed by the gauge-fixing condition to the identity everywhere except on a Dirac volume (bounded by the thin vortex sheet), where it has the effect of simply multiplying a certain set of links by -1. Upon center projection, $\widetilde{U}_\mu(x) \rightarrow Z_\mu(x)$ and $\widetilde{U}' \rightarrow Z'_\mu(x)$, and the projected configurations differ by the same discontinuous Z_2 gauge transformation. This discontinuity then shows up as an additional P-vortex in Z', not present in Z, at the location of the vortex inserted by hand.

While encouraging, this argument has two weaknesses: First, the inserted vortex is thin. The vortices which we are looking for in typical lattice configurations are thick. Secondly, the argument that U_A and U'_A transform into a unique configuration $\widetilde{U}_A$ in the adjoint representation ignores the fact that there exist a vast number of local maxima of the gauge-fixing functional R in (6.19), known as "Gribov copies." Existing numerical techniques are only able to fix the lattice to one of these local maxima, not to the global maximum. Despite this fact, the method has been shown, in numerical tests, to successfully locate thin vortices that were inserted by hand into the lattice [9].

The ability to find thin vortices inserted in the lattice is not unique to maximal center gauge, according to the above argument, but is shared by any choice of gauge which is defined as a condition on link variables in the adjoint representation. It is therefore worthwhile to explore some other possible gauge choices.

6.2.3 Indirect Maximal Center Gauge

This gauge [10] is useful in exploring connections between abelian monopoles and vortices.

The starting point is *maximal abelian gauge* [11], which minimizes the off-diagonal elements of the link variables, leaving a residual $U(1)^{N-1}$ gauge invariance (we will have more to say about this gauge, and its motivation, in Sect. 8.4). For $SU(2)$, maximize

$$R = \sum_{x,\mu} \text{Tr}[U_\mu(x)\sigma_3 U_\mu^\dagger(x)\sigma_3] . \tag{6.25}$$

This is equivalent to maximizing the (33) matrix element of the $SU(2)$ link variables in the adjoint representation. The link variables are decomposed as

$$U_\mu(x) = C_\mu(x)D_\mu(x) , \tag{6.26}$$

where D is the diagonal part of the link variable, rescaled to restore unitarity

$$D_\mu = \frac{1}{\sqrt{\left|[U_\mu]_{11}\right|^2 + \left|[U_\mu]_{22}\right|^2}} \begin{bmatrix} [U_\mu]_{11} & 0 \\ 0 & [U_\mu]_{22} \end{bmatrix}$$

$$= \begin{bmatrix} e^{i\theta_\mu} & 0 \\ 0 & e^{-i\theta_\mu} \end{bmatrix} . \tag{6.27}$$

Then we use the residual $U(1)$ symmetry to maximize

$$\tilde{R} = \sum_{x,\mu} |\mathrm{Tr}[D_\mu(x)]|^2 = \sum_{x,\mu} 4\cos^2\left(\theta_\mu(x)\right) , \tag{6.28}$$

leaving a residual center (in this case Z_2) gauge symmetry. This brings the configuration into the indirect maximal center gauge.

Both the direct and indirect maximal center gauges have Gribov copies: for any lattice configuration there are a huge number of local maxima of R (much like spin glasses [12]), and there is no known technique for finding the global maximum. There are two strategies for dealing with this problem. One can just make an effort to find the "best" copy one can, e.g. by a simulated annealing technique, or else give up on trying to find a best copy, and average over all Gribov copies. In Monte Carlo simulations this last strategy is easy; it is equivalent to simply picking a gauge copy at random.

6.2.4 Laplacian Center Gauge

Laplacian gauges, originally introduced by Vink and Wiese [13], are gauges which avoid the Gribov problem. Consider a Yang-Mills theory with two scalar fields ϕ_1^c, ϕ_2^c in the adjoint representation, where c is the color index. The unitary gauge

$$\phi_1^c(x) = \rho(x)\delta^{c3}$$

$$\phi_2^c(x) = 0 , \quad \phi_2^1(x) > 0 \tag{6.29}$$

can be fixed uniquely, leaving a residual Z_2 symmetry. In the Laplacian center gauge [14, 15], the two scalar fields are taken to be the two lowest eigenmodes

$$\sum_y -\Delta^{ab}(x, y) f_n^b(y) = \lambda_n f_n^a(x) \tag{6.30}$$

of the (negative) covariant Laplacian $-\Delta$ in adjoint representation

$$-\Delta^{ab}(x,y) = -\sum_{\mu} \left([U_{A\mu}(x)]^{ab}\delta_{y,x+\hat{\mu}} + [U_{A\mu}(x-\hat{\mu})]_{ba}\delta_{y,x-\hat{\mu}} - 2\delta_{xy}\delta^{ab} \right).$$

(6.31)

The Laplacian center gauge is the unitary gauge defined by (6.29), with the two lowest eigenmodes serving as the two scalar fields $\phi_{1,2}$.

6.2.5 Direct Laplacian Center Gauge

This is a hybrid [16], which for $SU(2)$ uses the two lowest eigenmodes of the adjoint covariant Laplacian to select a particular Gribov copy of direct maximal center gauge. In practice, the results obtained in this gauge are indistinguishable from picking gauge copies at random in direct maximal center gauge.

In the end, the only real justifications for using one or another gauge are empirical. Most of the results I will show were obtained in the direct Laplacian center gauge, but the corresponding results obtained in direct and indirect maximal center gauges are very similar.

6.3 The Numerical Evidence

The numerical results which bear on the center vortex mechanism fall into several categories:

1. Center Dominance: What string tension is obtained from P-vortices?
2. Vortex-Limited Wilson loops: What is the correlation between center-projected loops and Wilson loops on the unprojected lattice? Do P-vortices really locate thick vortices on the original lattice?
3. Vortex Removal: What is the effect of removing vortices, identified by center projection, from the lattice configuration?
4. Scaling: Does the density of vortices scale according to the asymptotic freedom prediction?
5. Finite Temperature
6. Chiral Condensate/Topological Charge

6.3.1 Center Dominance

The very first question is whether, under the factorization in one of the center gauges

$$U_{\mu}(x) = Z_{\mu}(x)V_{\mu}(x),$$

(6.32)

the variables $Z_\mu(x)$ of the center-projected configuration carry the disorder responsible for confinement.

The expectation value of a rectangular $R \times T$ Wilson loop typically has the following dependence on R and T:

$$W[R, T] = \exp[-(\sigma RT + \mu(R + T) - a(T/R + R/T) + b] . \qquad (6.33)$$

The term $\mu(R + T)$ is a self-energy term, and is divergent in the continuum limit. Creutz noticed that one could form a ratio of rectangular loops[2]

$$\chi[R, T] \equiv -\log \left\{ \frac{W[R, T]W[R - 1, T - 1]}{W[R - 1, T]W[R, T - 1]} \right\} , \qquad (6.34)$$

such that the self-energy terms cancel out, and in the limit of large loop areas the Creutz ratio converges to the asymptotic string tension

$$\chi[R, T] \to \sigma . \qquad (6.35)$$

Let $W_{cp}[R, T]$ be the Wilson loops of the center-projected lattice, i.e. loops constructed from the link variables $Z_\mu(x)$ identified in maximal center gauge, and let $\chi_{cp}[R, T]$ be the corresponding Creutz ratios. Then the question is whether, at large R, T, these Creutz ratios converge to the asymptotic string tension of the unprojected lattice.

Figure 6.3 is a first look at $\chi_{cp}[R, R]$ vs. coupling β for $R = 1$–6. There are three points, in particular, worth noting:

- At each β, the $\chi_{cp}[R, R]$ are almost identical for all $R > 1$. In other words, the asymptotic string tension is obtained already at $R = 2$.
- There is excellent agreement with asymptotic freedom, indicated by the straight line passing through the data points at $R = 2$. The slope of this line is a prediction of the Yang-Mills two-loop beta function.
- Even the data for $\chi_{cp}[1, 1]$ seems to fit with asymptotic freedom.

Creutz ratios in the full theory, by contrast, certainly do depend on R, and only converge to the asymptotic string tension, at a given β, at large R.

Figure 6.4 displays the (nearly constant) values for the Creutz ratios $\chi_{cp}[R, R]$ vs. R at several β values. At each β the central straight line represents the accepted value of the asymptotic string tension of lattice Yang-Mills theory; the upper and lower straight lines indicate the associated statistical errors. The agreement of the string tension on the center-projected lattice with the full string tension is known as *"center dominance."* The fact that $\chi_{cp}[R, R]$ is nearly R-independent means that

---

[2]The Creutz ratios $\chi[R, T]$ should not be confused with the group characters $\chi_r[g]$ discussed previously.

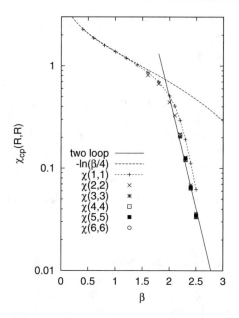

Fig. 6.3 Creutz ratios of center-projected Wilson loops. From Faber et al. [16]

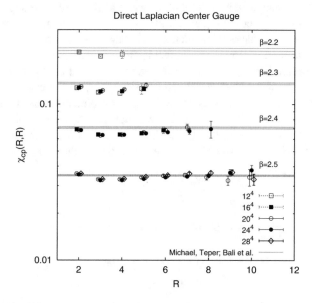

Fig. 6.4 Combined data, at $\beta = 2.2 - 2.5$, for center-projected Creutz ratios obtained after direct Laplacian center gauge fixing. Horizontal bands indicate the accepted values of asymptotic string tensions on the unprojected lattice, with the corresponding errorbars [17]. From Faber et al. [16]

the center-projected potential is linear starting from $R = 2$. This feature is known as *"precocious linearity."*

In order to understand precocious linearity, recall that center vortices on the unprojected lattice are thick objects, and the full effect on a Wilson loop, i.e. multiplication by a center element, is only obtained for thick vortices linked to large loops. Center projection essentially "shrinks" the thickness of the vortex to one lattice spacing, so the full effect of linking to a vortex appears for even the smallest center-projected Wilson loops. Thus, if P-vortex plaquettes are completely uncorrelated in a plane, then we must see a linear potential from the smallest distances. If this is not seen, then either the vortex surface is very rough, bending in and out of the plane, or else there are very small vortices. In either case there are correlations between nearby P-plaquettes, and a delay in the onset of the linear projected potential.

6.3.2   Vortex-Limited Wilson Loops

The fact that P-vortices in the projected lattice reproduce the correct, or nearly correct, asymptotic string tension does not, by itself, prove that these objects locate something physical on the unprojected lattice. P-vortices, after all, are identified with the help of gauge fixing; perhaps they are artifacts of the gauge choice. To address that issue, the crucial question to ask is whether P-vortices are correlated with gauge-invariant observables, such as ordinary Wilson loops.

A "vortex-limited Wilson loop" $W_n(C)$ is the expectation value of a Wilson loop on the unprojected lattice, evaluated in the subensemble of configurations in which the minimal area of the loop is pierced by precisely n P-vortices (i.e. there are n P-plaquettes in the minimal area). Here the center projection is used only to select the data set; the loop itself is evaluated using unprojected link variables.

If P-vortices on the projected lattice locate center vortices on the unprojected lattice, then for $SU(2)$ we would expect, asymptotically, that

$$\frac{W_n(C)}{W_0(C)} \to (-1)^n . \tag{6.36}$$

The reason is as follows: In maximal center gauge we have

$$W_n(C) = \langle Z(C) \text{Tr}[V(C)] \rangle_{n_p(C)=n}$$
$$= (-1)^n \langle \text{Tr}[V(C)] \rangle_{n_p(C)=n} , \tag{6.37}$$

where the $\langle \ldots \rangle_{n_p(C)}$ notation indicates that the planar loop is only evaluated in those configurations for which $n_p(C)$ P-plaquettes lie in the interior of the minimal area of the loop. If we assume that $V_\mu(x)$ has only short range correlations, then on a large loop this variable is insensitive to the presence or absence of vortices deep in

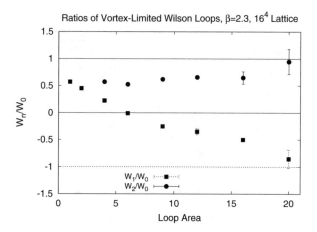

Fig. 6.5 W_n/W_0 ratios at $\beta = 2.3$ vs. loop area. Note that $W_1/W_0 \approx -1$, and $W_2/W_0 \approx 1$ for the largest loops. From Faber et al. [16]

the interior of the loop, i.e.

$$\langle \mathrm{Tr}[V(C)]\rangle_{n_p(C)=n} \approx \langle \mathrm{Tr}[V(C)]\rangle_{n_p(C)=0} \tag{6.38}$$

for large loops. From this, (6.36) follows immediately.

Figure 6.5 shows results for W_1/W_0, and W_2/W_0 for square $R \times R$ and rectangular $R \times (R+1)$ loops. The data is quite consistent with (6.36). One can also look at loops with even or odd numbers of P-vortices piercing the loop. We find, for $SU(2)$ [16]

$$\frac{W_{odd}(C)}{W_{even}(C)} \to -1 . \tag{6.39}$$

From the fact that $W_n/W_0 \to (-1)^n$, we conclude that P-vortices are correlated with the sign of the Wilson loop, in just the way expected if these P-vortices are correlated with the location of thick center vortices. If we restrict ourselves to loops with $n = 0$, or $n =$even, then there are no fluctuations in the sign of the loop due to vortices, and in the vortex scenario the asymptotic string tension should vanish. In fact, Creutz ratios constructed from loops of this kind show exactly the expected behavior, i.e.

$$\chi_0(R, R) \to 0 \quad , \quad \chi_{even}(R, R) \to 0 . \tag{6.40}$$

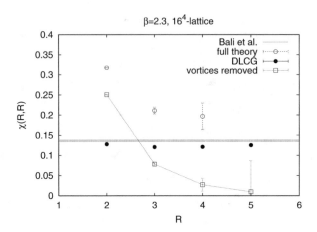

Fig. 6.6 Creutz ratios at $\beta = 2.3$ in the unprojected configuration (open circle), in the center projected configurations derived from direct Laplacian gauge (filled circle), and in the unprojected configurations with vortices removed (open square). The horizontal band indicates the accepted asymptotic string tension of the unprojected theory. From Faber et al. [16]

6.3.3   Vortex Removal

A powerful consistency test was devised by de Forcrand and D'Elia in [18]. Suppose we "remove" center vortices from the unprojected configuration, by multiplying a lattice configuration, in a center gauge, by its own center projection, i.e.

$$U_{\mu}(x) \rightarrow U'_{\mu}(x) = Z^*_{\mu}(x)U_{\mu}(x) = V_{\mu}(x) . \tag{6.41}$$

The center-projection of $U'_{\mu}(x)$ has no center vortices at all. What the procedure actually does is to insert a thin vortex in the middle of a thick vortex. The asymptotic fields of the thin and thick vortices would cancel out, removing the vortex disordering effect on large loops. Thus, if (1) P-vortices locate thick vortices (the evidence for that comes from the vortex-limited Wilson loops); and (2) vortex disorder is confining disorder (the evidence is center dominance); then removing vortices in this way should also remove the asymptotic string tension. This is, in fact, exactly what happens, as can be seen in Fig. 6.6.

6.3.4   Scaling of the P-vortex Density

If center vortices are physical objects, it makes sense that their density (vortex area per unit volume) in the vacuum is lattice-spacing independent in the continuum limit. If P-vortices lie in the middle of center vortices, it would likewise follow that P-vortex density is lattice-spacing independent, a property which was first checked in [6]. Let

$$N_{vor} = \text{total no. of P-plaquettes} = \text{total P-vortex area in lattice units}$$

$$N_T = \text{total no. of plaquettes} = \text{total lattice volume} \times 6 \;. \tag{6.42}$$

Then the density p of P-plaquettes is related to the vortex density ρ in physical units via

$$p = \frac{N_{vor}}{N_T} = \frac{N_{vor} a^2}{N_T a^4} a^2$$

$$= \frac{\text{Total Vortex Area}}{6 \times \text{Total Lattice Volume}} a^2$$

$$= \frac{1}{6}\rho a^2 \;. \tag{6.43}$$

If ρ is a physical quantity (i.e. β-independent), then we can substitute the asymptotic freedom expression for lattice spacing $a(\beta, \Lambda)$ to obtain the prediction

$$p = \frac{\rho}{6\Lambda^2}\left(\frac{6\pi^2}{11}\beta\right)^{102/121} \exp\left[-\frac{6\pi^2}{11}\beta\right] \;. \tag{6.44}$$

The average value of p is obtained from $W_{cp}[1, 1]$, because

$$\tfrac{1}{2}W_{cp}[1, 1] = (1 - p) + p \times (-1) = 1 - 2p \tag{6.45}$$

Figure 6.7 displays p vs. β; the straight line has the slope predicted from asymptotic freedom. In Fig. 6.8 we show closely related results by Gubarev et al.

Fig. 6.7 Evidence of asymptotic scaling of the P-vortex surface density. The data points are the number of P-plaquettes per unit volume, and the solid line is the asymptotic freedom prediction with $\sqrt{\rho/6\Lambda^2} = 50$. From Faber et al. [16]

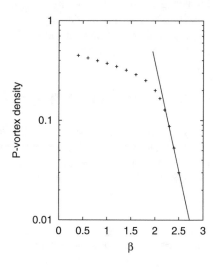

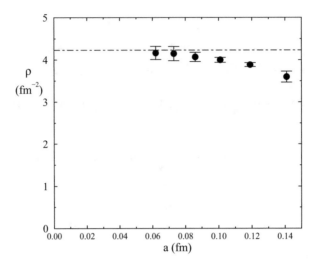

Fig. 6.8 P-vortex density vs. lattice spacing. From Gubarev et al. [19]

[19] using the indirect maximal center gauge. This is a plot of ρ in physical units vs. lattice spacing, and the insensitivity to lattice spacing at small lattice spacings is quite evident.

Gubarev et al. [19] also investigated the (gauge-invariant) action density, above the vacuum average, at the location of P-plaquettes, i.e.

$$\Delta S = -\tfrac{1}{2}\beta(W_1[1,1] - W[1,1]) . \tag{6.46}$$

It was found that this action density is singular in the continuum limit. The result means that in terms of action density, the middle of a center vortex stands out from the background, in the continuum limit, as a surface (or "brane") of infinite action density. The fact that the density of vortices is constant when taking the continuum limit, despite the infinite action density limit at the location of P-vortices, suggests a delicate cancellation between surface action and surface entropy.[3]

6.3.5 Vortices at High Temperatures

We have already encountered Polyakov loops, which measure the quark free energy in a periodic lattice at finite temperature $T = 1/(L_t a)$, where L_t is the extension of the lattice in the time direction. As the temperature is raised to T_c=220 MeV, quantized $SU(2)$ gauge fields theory go through a "deconfinement" transition,

---

[3]Zakharov [20] has argued that surfaces with singular action density fit in well with the need for certain power corrections in QCD sum rules.

where hadrons dissolve into their constituents. One can show numerically that at $T > T_c$ the quark free energy, as measured by the Polyakov loop, becomes finite. On the other hand, the vacuum of the deconfined phase still retains some features associated with the confined phase, since it has also been shown that spacelike Wilson loops (which are a measure of vacuum fluctuations) retain an area law and asymptotic string tension beyond the phase transition, even though the static quark potential measured by Polyakov loop correlators goes flat. A theory of confinement must be consistent with both of these features, i.e. an area law for spacelike loops but non-zero Polyakov loop expectation values; at $T > T_c$.

Center vortices fit in nicely with these requirements, and provide an intuitive picture of the transition which can be checked via numerical simulation. It is easiest to visualize the situation in 2+1 dimensions. Let us consider the expectation value of the product of two Polyakov loops, $P(\mathbf{x})P^*(\mathbf{x} + \mathbf{R})$. By the same argument as for a single Polyakov line, this quantity is related to the free energy of a quark-antiquark pair

$$e^{-\beta F_{q\bar{q}}} = \langle P(\mathbf{x})P^*(\mathbf{x} + \mathbf{R})\rangle \, , \tag{6.47}$$

where $\beta = 1/T$. In the confinement phase a flux tube, with some temperature-dependent string tension, will form between the quark-antiquark pair, and at large R the free energy of a separated quark-antiquark pair will approximate $\sigma(T)R + V_0$. This means that, in lattice units,

$$\langle P(\mathbf{x})P^*(\mathbf{x} + \mathbf{R})\rangle \propto e^{-\sigma(T)L_t R} \, . \tag{6.48}$$

In the vortex theory, this exponential falloff, proportional to the minimal area bounded by the Polyakov loops, is due to random (uncorrelated) piercings of the minimal area by vortices. However, vortices have a certain thickness, and it is known from the work of Kovacs and Tomboulis [21], and de Forcrand and von Smekal [22], that when vortices are "'squeezed" by the finite extension of the lattice, i.e. the thickness of the lattice is less than the normal thickness of center vortices, then the free energy of the squeezed vortices rises rapidly away from zero as the lattice extension is reduced, and at some point squeezed vortices cease to percolate. The situation in 2+1 dimensions is sketched in Fig. 6.9. When the squeezed vortices no longer percolate, then piercings of the minimal surface between Polyakov loops are correlated, and the piercings can no longer result in the exponential falloff shown in (6.48). An exponential falloff of the correlator guarantees that $\langle P(x)\rangle = 0$, but in the absence of this falloff there is no reason for the expectation value of the Polyakov loop to vanish, and we have

$$\begin{aligned}
e^{-\beta F_{q\bar{q}}} &= \langle P(\mathbf{x})P^*(\mathbf{x} + \mathbf{R})\rangle \\
&\rightarrow \langle P(\mathbf{x})\rangle\langle P^*(\mathbf{x} + \mathbf{R})\rangle \\
&> 0 \, .
\end{aligned} \tag{6.49}$$

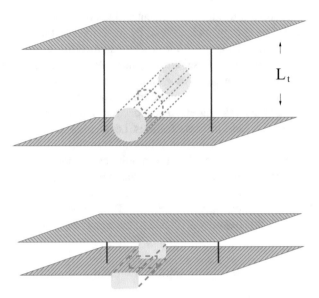

Fig. 6.9 Vortices running in the spacelike directions disorder Polyakov loops. When the time extension L_t is smaller than the diameter of the vortex (high temperature case), then spacelike vortices are "squeezed" and cease to percolate

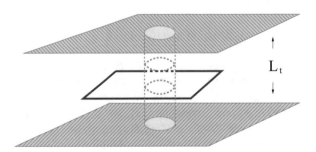

Fig. 6.10 Vortices running in the timelike direction disorder spacelike Wilson loops. The vortex cross section is not constrained by a small extension in the time direction

Since the "cost" in free energy of having an infinitely separated quark and antiquark is finite, this is the deconfined phase.

On the other hand, let us consider spacelike loops at high temperature. The area law for these loops is due to piercing by vortices oriented in the time direction, as indicated in Fig. 6.10. However, the thickness of such vortices is not constrained by the finite lattice extension in the time direction, and their free energy on a lattice with large extension in all space directions is essentially zero. It is necessary, however, that these unsuppressed configurations are oriented in the time direction all along their length, which means such vortices will extend through the lattice in the time

Fig. 6.11 Schematic picture of P-vortices in a "space-slice" of the lattice, below and above the deconfinement phase transition. From Engelhardt et al. [8]

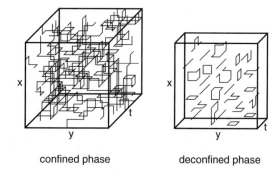

confined phase   deconfined phase

direction, and close via the lattice periodicity. Vortices of this type account for the area law falloff of spatial Wilson loops, even at very high temperatures.

In $3 + 1$ dimensions the vortex lines become vortex sheets winding through the periodic lattice in the time direction, but to visualize the situation it is helpful to consider a space-slice of the lattice, say at some fixed value of the z-coordinate. In a space-slice the vortex surfaces are again line-like. The P-vortices which we use to locate center vortices have, of course, no thickness, and are literally lines. The distribution of P-vortices in a space-slice is shown schematically in Fig. 6.11. At low temperatures, the P-vortices percolate throughout the lattice, while at high temperatures the vortex loops are small, but some of them (the ones responsible for the spacelike string tension) wind through the lattice in the periodic time direction.

On a time-slice, in $D = 2 + 1$ dimensions, the P-vortices would simply be a lot of uncorrelated points (P-plaquettes) in the plane. In $D = 3 + 1$ dimensions what we should see is the cross-section of P-vortex surfaces which are closed by lattice periodicity in the time direction, but percolate in spatial directions. Thus, in a time slice, we would expect to see large loops percolating throughout the spatial volume, even at high temperatures.

This picture and its numerical verification are due to the Tübingen group in [8] (see also refs. [23, 24]). Figure 6.12 shows some data for a space-slice at finite temperature. The x-axis is in units of the maximal extension of the vortex loop in the $L^2 \times L_t$ space-slice 3-volume, i.e. $\sqrt{2(L/2)^2 + (L_t/2)^2}$. The y-axis is the fraction of P-vortex plaquettes in the space-slice which are dual to a vortex loop of a given extension. At low temperature, most P-plaquettes belong to vortex loops whose extensions are comparable to the size of the lattice. At high temperatures, most P-plaquettes belong to short loops in the space-slice. The same sort of data for a time slice is shown in Fig. 6.13, at temperatures just below, just above, and well above the deconfinement transition. On a time slice there is little difference with temperature. At any temperature, most P-plaquettes belong to P-vortices whose extension is comparable to the size of the lattice. Thus we have percolation of vortices in a time-slice, and a corresponding area-law falloff for spacelike Wilson loops.

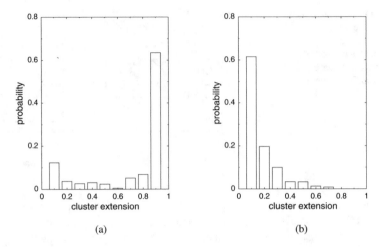

(a)   (b)

Fig. 6.12 Histograms of vortex extension in a space-slice at finite temperature, both below (left figure (**a**), $T = 0.7T_c$) and above (right figure (**b**), $T = 1.85T_c$) the deconfinement phase transition. The data is for $\beta = 2.4$ on a $12^3 \times 7$ (confined) and $12^3 \times 3$ (deconfined) lattices; center projection in direct maximal center gauge. From Engelhardt et al. [8]

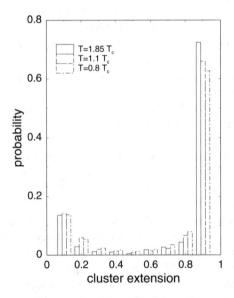

Fig. 6.13 Histograms of vortex extension in a time slice at finite temperature. Vortices are identified by center projection at $\beta = 2.4$ in direct maximal center gauge. Data at three temperatures are shown on the same figure. From Engelhardt et al. [8]

6.3.6   Chiral Condensates and Topological Charge

Consider a gauge theory with two light quarks, u and d, which together form a flavor doublet. As long as the masses of these quarks are zero, the Lagrangian is invariant under the transformations

$$\begin{bmatrix} u \\ d \end{bmatrix} \rightarrow U_B \times U_A \times U_L \times U_R \begin{bmatrix} u \\ d \end{bmatrix}, \tag{6.50}$$

where

$$U_B = \exp[i\theta_B]$$

$$U_A = \exp[i\theta_A \gamma_5]$$

$$U_L = \exp[i\theta_L^i (1 - \gamma_5)\frac{\sigma_i}{2}]$$

$$U_R = \exp[i\theta_R^i (1 + \gamma_5)\frac{\sigma_i}{2}]. \tag{6.51}$$

The symmetry with respect to $U(1)$ transformations U_B results, via Noether's Theorem, in conservation of baryon number. The axial $U(1)$ symmetry with respect to transformations U_A, while a symmetry of the Lagrangian, is not a symmetry of the full quantum theory. This symmetry is said to be *anomalous*. One way of understanding this anomaly in the continuum, in the context of path-integral quantization, is that the path-integration measure for the fermions is not invariant under U_A [25].[4] With most lattice regularizations the lattice measure is invariant under the axial $U(1)$ symmetry, and the symmetry is instead broken by the regularized lattice action [27]. Finally, while the chiral $SU(2)_L \times SU(2)_R$ symmetry transformations $U_L \times U_R$ are not broken by any anomaly, neither is this symmetry realized in the spectrum of the theory. If it were, then particles would appear in equal mass doublets of opposite parity, and this is something which is not found in nature. Instead, chiral symmetry is spontaneously broken to ordinary $SU(2)$ isospin symmetry, which consists of the subgroup of transformations for which $\theta_L^i = \theta_R^i$. When a continuous global symmetry is broken spontaneously, the Goldstone theorem assures us that there must be some associated massless particles; in this case these are the pions, which would be massless if the quarks were exactly massless.[5]

---

[4] Another picture, in the Hamiltonian formulation, involves the notion of the Dirac sea. Under the influence of an external gauge field, the energy levels of the Dirac operator shift up or down, depending on the chirality of the state. This level-shifting can *appear* as the creation of quark-antiquark pairs by a gauge field, with quantum numbers violating conservation of the axial $U(1)$ symmetry [26].

[5] For a theory with N_f massless flavors of quarks, the chiral symmetry is $SU(N_f)_R \times SU(N_f)_L$, which is spontaneously broken to an $SU(N_f)$ flavor symmetry, and the massless Goldstone

Whenever there is a possibility of spontaneously broken symmetry, it is useful to introduce a local operator which is not invariant under the symmetry, and whose expectation value would average to zero if the symmetry were unbroken. The expectation value of such an operator is the order parameter for the symmetry breaking. For the Ising model the operator was the average spin, whose expectation value is the magnetization, and this is non-zero if the Z_2 global symmetry of the Ising model is spontaneously broken. In the case of global center symmetry, the Polyakov loop is the natural order parameter. For chiral symmetry, an appropriate order parameter is $\langle \overline{\psi} \psi \rangle$, and the symmetry is broken when this quantity is non-zero. This order parameter, known as the *chiral condensate*, is also of phenomenological importance. If the quark masses (known as *current* quark masses) are not exactly zero, then the pions, now described as "pseudo-Goldstone" particles, also acquire a mass, and this mass can be expressed in terms of the pion decay constant f_π, the quark masses, and the order parameter for chiral symmetry breaking:

$$m_\pi^2 = -2 \frac{m_u + m_d}{f_\pi^2} \langle \overline{\psi} \psi \rangle \ . \tag{6.52}$$

A derivation can be found in standard texts, e.g. [28].

As we saw in the case of the Ising model, a non-zero value for the order parameter can only be obtained via a careful order of limits: First we introduce a term which breaks the symmetry explicitly (an external magnetic field, for the Ising model, or a quark mass term, in the present case), take the infinite volume limit, and only then remove the explicit breaking term. Proceeding in this way, Banks and Casher [29] found a beautiful relationship between the chiral condensate and the density of near-zero eigenvalues of the Dirac operator. The derivation is as follows: Introduce a small quark mass m, which is to be taken to zero at the end of the calculation, and let

$$i \, \slashed{D} \phi_n = \lambda_n \phi_n \tag{6.53}$$

be the eigenvalue equation for the Euclidean-space Dirac operator, with $\slashed{D} = \gamma^\mu D_\mu$. Then the massive quark propagator can be expressed as

$$\langle \psi(x) \overline{\psi}(y) \rangle = \sum_n \left\langle \frac{\phi_n(x) \phi^\dagger(y)}{-i \lambda_n + m} \right\rangle \ . \tag{6.54}$$

---

particles belong to a multiplet transforming in the adjoint representation of the remaining flavor symmetry.

We set $x = y$ to obtain the chiral condensate, then integrate over x (noting the orthonormality of the ϕ_n), and divide by spacetime volume, which gives

$$\langle \overline{\psi}\psi \rangle = -\frac{1}{V}\left\langle \sum_n \frac{i\lambda_n + m}{\lambda_n^2 + m^2}\right\rangle$$

$$= -\frac{2m}{V}\left\langle \sum_{\lambda_n>0} \frac{1}{\lambda_n^2 + m^2}\right\rangle , \tag{6.55}$$

where we have used the fact that, apart from zero modes, the $\{i\lambda_n\}$ come in complex-conjugate pairs, so the term with λ_n in the numerator averages to zero. Next take the infinite volume limit, where the Dirac operator has a continuous spectrum, and

$$\frac{1}{V}\left\langle \sum_{\lambda_n>0} \dots \right\rangle \rightarrow \int d\lambda\; \rho(\lambda)\langle\dots\rangle , \tag{6.56}$$

where $\rho(\lambda)$ is the density of eigenvalues per unit volume. This gives

$$\langle \overline{\psi}\psi \rangle = -2m \int d\lambda\; \frac{\rho(\lambda)}{\lambda^2 + m^2} . \tag{6.57}$$

The infinite volume limit is followed by the zero mass limit, using $m/(\lambda^2 + m^2) \rightarrow \pi\delta(\lambda)$, from which we finally obtain the Banks-Casher formula

$$\langle \overline{\psi}\psi \rangle = -\pi\rho(0) . \tag{6.58}$$

From the point of view of the Banks-Casher formula, the mechanism of chiral symmetry breaking is tied to whatever property of vacuum field configurations gives rise to $\rho(0) > 0$. Note that $\rho(0) = 0$ at the level of perturbation theory. For example, the eigenvalues of the ordinary Laplacian operator $-\nabla^2$ are $\lambda = p^2$. Then, from

$$\int d^D p = \int d\Omega \int_0^\infty dp\; p^{D-1} \propto \int_0^\infty d\lambda\; \lambda^{(D-2)/2} , \tag{6.59}$$

we read off that in this case $\rho(\lambda) \propto \lambda^{(D-2)/2}$ in D Euclidean spacetime dimensions, so that $\rho(0) = 0$ for any $D > 2$. A similar argument holds for the density of states of the free Dirac operator.

Chiral symmetry breaking clearly does not require a non-vanishing asymptotic string tension, since the asymptotic string tension in real QCD, with light quarks, vanishes due to the string breaking mechanism. Chiral symmetry breaking does not even require gauge fields. The breaking can occur in theories with other types of forces between fermions, as in the Nambu Jona-Lasinio model [30], where the effect is due to four-fermion interactions. Nor is chiral symmetry breaking a necessary feature of asymptotically free gauge field theories; some counterexamples in supersymmetric confinement-like theories are known [31]. But for present purposes it is sufficient to note that chiral symmetry breaking certainly does occur in QCD, and $\langle \bar{\psi} \psi \rangle$, is perfectly well-defined even in a theory without dynamical fermions: it is simply the trace of the quark propagator evaluated in a pure gauge theory.[6] It is entirely possible that the mechanism for chiral symmetry breaking in non-abelian gauge theories is not so closely related to the confinement mechanism. But it would be more economical, and more satisfying, if the objects which we suspect are responsible for confinement, i.e. the center vortices, are also somehow essential for a non-zero value for the chiral condensate.

On the lattice the chiral condensate can be computed directly, and it is also possible to numerically compute the low-lying spectrum of the Dirac operator. However, the lattice regularization of this operator introduces some thorny technical issues, which are discussed in most modern texts on lattice gauge theory (cf. in particular, Smit [32] and DeGrand and DeTar [33]). Briefly, the problem is that a naive lattice discretization, simply replacing covariant derivatives in the continuum by the one-link lattice covariant derivatives, leads to the notorious doubling problem: instead of one light fermion, one actually ends up with 2^D light fermions on a hypercubic lattice. One method of handling this problem, by introducing a one link mass term with a carefully tuned coefficient, breaks the chiral symmetry explicitly, at least at finite lattice spacing. Another formulation, known as *staggered fermions*, which places different fermionic components on neighboring lattice sites, reduces the multiplicity to four (known as "tastes") in $D = 4$ dimensions, while preserving a subgroup of the chiral symmetry. A variation of staggered fermions, known as *asqtad* fermions, introduces some next-to-nearest neighbor operators in order to reduce lattice artifacts, and improve convergence to the continuum limit. It has been used extensively in numerical simulations. Finally, the *overlap* Dirac operator has no doubling problem whatever, and is endowed with an exact invariance which, while not exactly the chiral symmetry shown in (6.51), approximates the standard $SU(2)_L \times SU(2)_R$ symmetry for a sufficiently smooth lattice gauge field background. We will not be detained by these issues here, except to mention that the results described below were obtained using either the staggered or the asqtad discretization of the Dirac operator. The overlap formulation is not so suitable for our purposes, because center-projected configurations are as far from smooth as possible, and the symmetry of the overlap operator is therefore not so close to the standard $SU(2)_L \times SU(2)_R$ chiral symmetry (cf. the discussion in [34]).

Lattice simulations by de Forcrand and co-workers [15, 18] have found that when vortices are removed via the procedure of Eq. (6.41), then the chiral condensate vanishes in the vortex-removed configuration, while in center-projected configurations, the value of the condensate is greatly enhanced, compared to its value in unprojected configurations. Their results are shown in Fig. 6.14. The low-lying spectrum of the

---

[6]It should be noted, however, that $\langle \bar{\psi} \psi \rangle$ diverges like $\log(m)$ in theories with no dynamical fermions, as $m \to 0$.

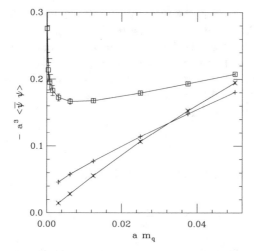

Fig. 6.14 Chiral symmetry breaking order parameter $\langle \bar{\psi} \psi \rangle$ vs. lattice bare quark mass m_q, for the unmodified (plus sign), center-projected (open square), and vortex removed (multiplication sign) configurations in pure $SU(2)$ lattice gauge theory. The order parameter for vortex removed configurations extrapolates to zero, as $m_q \to 0$. From Alexandrou, de Forcrand and D'Elia [15]

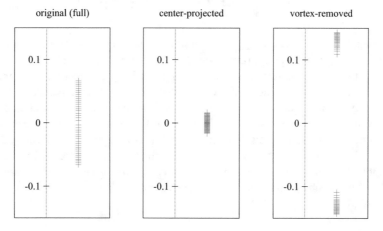

Fig. 6.15 Spectrum of the lowest-lying eigenvalues of the lattice Dirac operator (asqtad formulation), for unmodified, center projected, and vortex-removed configurations, in a pure SU(2) gauge theory. See Hollwieser et al. [34] for details of the simulation

Dirac operator in center-projected and vortex-removed configurations has also been calculated, in ref. [34], and the results for the first 20 complex-conjugate pairs of smallest-magnitude Dirac eigenvalues are shown in Fig. 6.15 (numerical details can be found in the cited reference). What is striking is that when vortices are removed, a large gap opens up around the zero eigenvalue; the eigenvalue density is clearly zero at $\lambda = 0$. In the center-projected configurations, the density of eigenvalues

appear to be larger than in the unmodified configurations. This is all consistent
with the Banks-Casher relation, and the results shown in Fig. 6.14. The evidence
is that center vortices are not only responsible for the confining force, but also, by
themselves, can be responsible for the spontaneous breaking of chiral symmetry.

Next, let us return to the axial $U(1)$ symmetry, which is broken at the quantum
level by an anomaly. In the absence of the anomaly, Noether's theorem would lead
to a conserved current, $\partial^\mu j_{5\mu} = 0$, where

$$j_{5\mu}(x) = \overline{\psi}(x)\gamma_\mu\gamma_5\psi(x) . \tag{6.60}$$

In fact the correct answer for massless quarks is [35]

$$\partial^\mu j_{5\mu} = -\frac{g^2 N_f}{16\pi^2}\varepsilon^{\mu\nu\alpha\beta}\mathrm{Tr}[F_{\mu\nu}F_{\alpha\beta}] , \tag{6.61}$$

where N_f is the number of light flavors (= 2, if we consider only a doublet of u, d
quarks). The integral of the right-hand side of this equation is proportional to a
quantity known as the *topological charge*

$$Q = \frac{1}{32\pi^2}\int d^4x \ \varepsilon^{\mu\nu\alpha\beta}\mathrm{Tr}[F_{\mu\nu}F_{\alpha\beta}] . \tag{6.62}$$

The integrand $\varepsilon^{\mu\nu\alpha\beta}\mathrm{Tr}[F_{\mu\nu}F_{\alpha\beta}]$ can be expressed as a total divergence. There are,
however, finite action configurations (such as instantons) for which the integral of
this quantity is non-vanishing, and Q is an integer [36]. The restriction to integer
values is of course no accident. There is a celebrated result in mathematics (known
as the Atiyah-Singer Index Theorem) which tells us that the topological charge of
any given gauge-field configuration is equal to the number of zero modes of the
Dirac operator of positive chirality (i.e. they are also eigenstates of the γ_5 Dirac
matrix with positive eigenvalue), minus the number of zero modes with negative
chirality; this difference is obviously an integer. The mean-square variation of
the topological charge, per unit spacetime volume, is known as the topological
susceptibility

$$\chi = \frac{\langle Q^2 \rangle}{V} , \tag{6.63}$$

and this quantity also plays an important role in phenomenology. If the axial
$U(1)$ symmetry were conserved but spontaneously broken, then there would be
an associated massless Goldstone boson (if the quarks were massless) or a light
pseudo-Goldstone boson (if the quarks had a small current mass). This would be
the η' boson, which at about 1 GeV is far too heavy to be considered a pseudo-
Goldstone particle. An important result due to Witten [37] and Veneziano [38]
relates the topological susceptibility χ, computed in a pure gauge theory, to the

η' mass

$$m_{\eta'}^2 = \frac{2N_f}{f_\pi^2} \chi \ , \tag{6.64}$$

and this is known, naturally enough, as the Witten-Veneziano mass formula.

The topological susceptibility, like the chiral condensate, is a property of the vacuum state, and has a well defined expectation value in a pure gauge theory with no quark fields whatever. It is not obvious that topological susceptibility has any direct connection to the ultra long-range property of confinement, or to center vortices. It could be that other non-confining configurations are responsible for the density of topological charge, and instantons (which carry integer amounts of topological charge) have been the traditional candidates. An indication that center vortices might play a role was the result, due to de Forcrand and D'Elia [18], that topological charge plummets to zero when vortices are removed from lattice configurations. A more detailed analysis is due to Engelhardt [39], who has pointed out that on a thin vortex surface, topological charge density can arise at sites on the surface where the tangent vectors to the surface are in all four space-time directions. These sites are of two sorts:

- Self-intersections, where two surface segments intersect at a point in four dimensions.
- "Writhing" points, in which the vortex surface twists about in such a way as to produce four linearly independent tangent vectors to the surface.

Writhing points are best illustrated rather than described, and both types of sites generating topological charge are indicated in Fig. 6.16. One can show that zero modes of the Dirac operator tend to peak at these intersection and writhing points; the plot in Fig. 6.17 shows the modulus of Dirac zero modes in a background of four intersecting P-vortex sheets.

Calculation of the topological charge of vortex surfaces still requires the assignment of an "orientation" for the field strength, and a suitable algorithm was

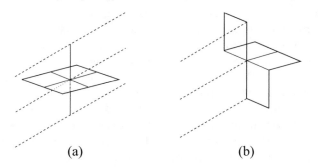

(a)   (b)

Fig. 6.16 Intersection points (**a**) and writhing points (**b**) which contribute to the topological charge of a P-vortex surface. From Reinhardt [40]

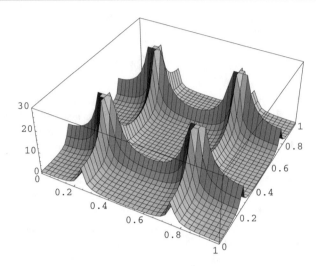

Fig. 6.17 Quark zero modes in a background of intersecting P-vortex sheets. From Reinhardt et al. [41]

proposed by Engelhardt in ref. [39]. Using this proposal, Bertle, Engelhardt, and Faber [42] have calculated the topological susceptibility derived from P-vortices, and have found results which are consistent with the conventional measurements of this quantity.

6.3.7 Center Symmetry Breaking by Matter Fields

We have so far only considered center vortices in theories with a center-symmetric action. But what happens in a theory where the center symmetry is explicitly broken by dynamical matter fields (such as quarks) in the fundamental representation of the gauge group? Theories of this sort are at most "confinement-like," in the sense that the static quark potential rises linearly only up to some string-breaking scale, which increases with the mass of the matter fields. Beyond that scale the string tension is zero. The question is whether the assumed dominance of large-scale vacuum fluctuations by center vortices can be consistent with this behavior.

This question is easiest to address numerically in a theory where the matter fields are scalar, rather than fermionic, such as the SU(2) gauge-Higgs theory introduced in Sect. 3.2, Eq. (3.6). We recall that the phase diagram of this theory, sketched in Fig. 3.4, has a line of sharp crossover behavior. The confinement-like region (with a string tension at intermediate scales) lies below this line, and the Higgs-like region above. In this theory we can locate vortices in the usual way, by gauge-fixing to direct maximal center gauge and then applying center projection. The aim is to test center dominance in the confinement-like region, but close enough to the crossover line so that the screening effect of the scalar field is detectable numerically. Screening is easiest to observe, in both the full and center-projected lattices, in the expectation values of Polyakov loops, and Polyakov loop correlators.

Center-projected Polyakov loops in the SU(2) gauge-Higgs theory were computed in [43]. At $\beta = 2.2$, the crossover to the Higgs-like region occurs at about $\gamma = 0.84$, while center symmetry is unbroken at $\gamma = 0$. In the latter case the expectation value of a Polyakov loop, like any order parameter for spontaneous symmetry breaking, is vanishing in a finite volume. It is standard practice to instead measure

$$\langle P \rangle \equiv \left\langle \frac{1}{L^3} \left| \sum_{\mathbf{x}} P(\mathbf{x}) \right| \right\rangle \tag{6.65}$$

where $P(\mathbf{x})$ denotes the Polyakov loop passing through the point $\{\mathbf{x}, t = 0\}$. In the case of unbroken center symmetry, at $\gamma = 0$ and on an $L^3 \times L_T$ lattice, we must find

$$\langle P \rangle \propto \sqrt{\frac{1}{L^3}} \tag{6.66}$$

The reason is that if Polyakov loops have a finite correlation length l, then they must fluctuate independently in regions of extension L/l, and the number of such regions is proportional to the spatial volume L^3. This means, by the usual laws of statistics, that the magnitude of the average value of P, in any large-volume lattice configuration, must be proportional to $L^{-3/2}$. For explicitly broken center symmetry ($\gamma \neq 0$) we must instead find that $\langle P \rangle$ has a non-zero limit at $L \to \infty$.

The data for center-projected Polyakov loops in the confinement-like region at $\beta = 2.2$ and $\gamma = 0.71$, and also in the center-symmetric case of $\gamma = 0$, are shown in Fig. 6.18a, for lattice sizes $L^3 \times 4$ up to $L = 20$. The straight line is

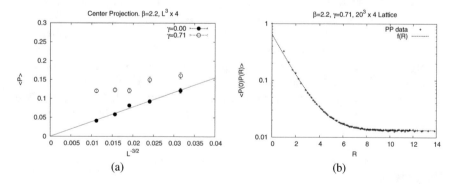

(a)   (b)

Fig. 6.18 (a) Center-projected Polyakov loop values in the SU(2) gauge-Higgs model at $\beta = 2.2$, $\gamma = 0$ and $\gamma = 0.71$, on $L^3 \times 4$ lattice volumes with $L = 10, 12, 14, 16, 20$ The straight line is a best fit to the $\gamma = 0$ data. (b) Polyakov loop correlator $\langle P(0)P(R) \rangle$ on the center-projected lattice. From [43]

a best fit through the $\gamma = 0$ data, and errorbars for some data points are smaller than the symbol size. It is clear that the $\gamma = 0$ data is consistent with Eq. (6.66), and $\langle P \rangle$ in center projection extrapolates to zero in the infinite volume limit. At $\gamma = 0.71$ the system is still below the line of sharp crossover behavior, and in the temporary confinement region. It appears from the data that at this coupling $\langle P \rangle$ has stabilized (at $L = 14, 16, 20$) to a non-zero value of $\langle P \rangle \approx 0.120(4)$. So at $\beta = 2.2$, $\gamma = 0.71$, color screening of Polyakov loops by the matter field is detectable. A similar result was obtained in [44], for the variable Higgs-modulus version of the theory.

Figure 6.18b displays the correlator of center-projected Polyakov loops $\langle P(x)P(x + R) \rangle$ at $\beta = 2.2$, $\gamma = 0.71$, on a $20^3 \times 4$ lattice. The dashed line is a best fit to the data, for $R \geq 2$, by the function

$$f(R) = c_0 + c_1 \exp[-4\sigma R] \tag{6.67}$$

From the fit we find $c_0 = 0.0182$, $\sigma = 0.211$. Not surprisingly, c_0 is close to the square of the expectation value of the Polyakov loop in center projection. The transition from a linear static quark potential, in the region where the correlator falls exponentially, to a constant potential, occurs at around five lattice spacings. This agrees with the corresponding results on the unprojected lattice. It was also found that in the Higgs-like region, the center-projected Polyakov-loop correlator shows no evidence of a linear potential at any length scale.

What we learn from this data is that in a confinement-like theory, where the center symmetry is broken explicitly by matter fields, center vortices do not simply disappear. In the SU(2) gauge-Higgs theory they appear to be responsible for the string tension at intermediate scales, while allowing for a vanishing string tension asymptotically. There is still, however, an unsettled question of exactly *how* the vortex structure in confinement-like theories differs asymptotically from the vortex structure in center-symmetric theories, so as to avoid generating an asymptotic string tension. Matter fields which break center symmetry explicitly will cause the Dirac volume associated with center vortices to acquire an action density. We therefore expect that on large scales the size of the Dirac volumes will be far smaller than in the center-symmetric theory. Since the vortices themselves do not generate an asymptotic string tension, vortex piercings of the minimal area of a Wilson loop must be correlated at large scales. It has been suggested [43] that the shrinking of the Dirac volume, and loss of an asymptotic string tension, could be achieved if the vortex surface has a branched-polymer structure at large scales, but at present this is only a speculation.

References

1. G. 't Hooft, On the phase transition towards permanent quark confinement. Nucl. Phys. B **138**, 1 (1978)
2. G. Mack, Properties of lattice gauge theory models at low temperatures, in *Recent Developments in Gauge Theories*, ed. by G. 't Hooft et al. (Plenum, New York, 1980)

3. H.B. Nielsen, P. Olesen, A quantum liquid model for the QCD vacuum: gauge and rotational invariance of domained and quantized homogeneous color fields. Nucl. Phys. B **160**, 380 (1979); J. Ambjørn, P. Olesen, A color magnetic vortex condensate in QCD. Nucl. Phys. B **170**, 60, 265 (1980)

4. J. Cornwall, Quark confinement and vortices in massive gauge invariant QCD. Nucl. Phys. B **157**, 392 (1979)

5. R. Feynman, The qualitative behavior of Yang-Mills theory in 2+1 dimensions. Nucl. Phys. B **188**, 479 (1981)

6. M. Engelhardt, K. Langfeld, H. Reinhardt, O. Tennert, Interaction of confining vortices in SU(2) lattice gauge theory. Phys. Lett. B **431**, 141 (1998). arXiv:hep-lat/9801030

7. L. Del Debbio, M. Faber, J. Giedt, J. Greensite, Š. Olejník, Detection of center vortices in the lattice Yang-Mills vacuum. Phys. Rev. D **58**, 094501 (1998). arXiv:hep-lat/9801027

8. M. Engelhardt, K. Langfeld, H. Reinhardt, O. Tennert, Deconfinement in SU(2) Yang-Mills theory as a center vortex percolation transition. Phys. Rev. D **61**, 054504 (2000). arXiv:hep-lat/9904004

9. M. Faber, J. Greensite, Š. Olejník, D. Yamada, The vortex-finding property of maximal center (and other) gauges. J. High Energy Phys. **9912**, 012 (1999). arXiv:hep-lat/9910033

10. L. Del Debbio, M. Faber, J. Greensite, Š. Olejník, Center dominance and Z(2) vortices in SU(2) lattice gauge theory. Phys. Rev. D **55**, 2298 (1997). arXiv:hep-lat/9610005

11. A. Kronfeld, M. Laursen, G. Schierholz, U.-J. Wiese, Monopole condensation and color confinement. Phys. Lett. B **198**, 516 (1987)

12. K.H. Fischer, J.A. Hertz, *Spin Glasses* (Cambridge University Press, Cambridge, 1991)

13. J. Vink, U.-J. Wiese, Gauge fixing on the lattice without ambiguity. Phys. Lett. B **289**, 122 (1992). arXiv:hep-lat/9206006; J. Vink, Investigation of Laplacian gauge fixing for U(1) and SU(2) gauge fields. Phys. Rev. D **51**, 1292 (1995). hep-lat/9407007

14. Ph. de Forcrand, M. Pepe, Center vortices and monopoles without lattice Gribov copies. Nucl. Phys. B **598**, 557 (2001). arXiv:hep-lat/0008016

15. C. Alexandrou, Ph. de Forcrand, M. D'Elia, The role of center vortices in QCD. Nucl. Phys. A **663**, 1031 (2000). arXiv:hep-lat/9909005

16. M. Faber, J. Greensite, Š. Olejník, Direct Laplacian center gauge. J. High Energy Phys. **11**, 053 (2001). arXiv:hep-lat/0106017

17. G. Bali, C. Schlichter, K. Schilling, Observing long color flux tubes in SU(2) lattice gauge theory. Phys. Rev. D **51**, 5165 (1995). arXiv:hep-lat/9409005; C. Michael, M. Teper, Towards the continuum limit of SU(2) lattice gauge theory. Phys. Lett. B **199**, 95 (1987)

18. Ph. de Forcrand, M. D'Elia, On the relevance of center vortices to QCD. Phys. Rev. Lett. **82**, 4582 (1999). arXiv:hep-lat/9901020

19. F.V. Gubarev, A.V. Kovalenko, M.I. Polikarpov, S.N. Syritsyn, V.I. Zakharov, Fine tuned vortices in lattice SU(2) gluodynamics. Phys. Lett. B **574**, 136 (2003). arXiv:hep-lat/0212003

20. V.I. Zakharov, Non-perturbative match of ultraviolet renormalon (2003). arXiv:hep-ph/0309178

21. T. Kovács, E. Tomboulis, Computation of the vortex free energy in SU(2) lattice gauge theory. Phys. Rev. Lett. **85**, 704 (2000). arXiv:hep-lat/0002004

22. L. von Smekal, Ph. de Forcrand, 't Hooft loops, electric flux sectors, and confinement in lattice Yang-Mills theory. Phys. Rev. D **66**, 011504 (2002). arXiv:hep-lat/0107018; L. von Smekal, Ph. de Forcrand, Electric and magnetic fluxes in SU(2) Yang-Mills theory. Nucl. Phys. Proc. Suppl. **119**, 655 (2005). arXiv:hep-lat/0209149; L. von Smekal, Ph. de Forcrand, O. Jahn, More on electric and magnetic fluxes in SU(2). arXiv:hep-lat/0212019

23. K. Langfeld, O. Tennert, M. Engelhardt, H. Reinhardt, Center vortices of SU(2) Yang-Mills theory at finite temperatures. Phys. Lett. B **452**, 301 (1999). arXiv:hep-lat/9805002

24. M. Chernodub, M. Polikarpov, A. Veselov, M. Zubkov, Center vortices and center monopoles in SU(2) lattice gluodynamics. Nucl. Phys. Proc. Suppl. **73**, 575 (1999). arXiv:hep-lat/9809158

25. K. Fujikawa, Path integral measure for gauge invariant fermion theories. Phys. Rev. Lett. **42**, 1195 (1979)

26. J. Ambjorn, J. Greensite, C. Peterson, The axial anomaly and the lattice Dirac sea. Nucl. Phys. B **221**, 381 (1983)
27. L.H. Karsten, J. Smit, Lattice fermions: species doubling, chiral invariance, and the triangle anomaly. Nucl. Phys. B **183**, 103 (1981)
28. S. Weinberg, *The Quantum Theory of Fields*, vol. II (Cambridge University Press, Cambridge, 2005)
29. T. Banks, A. Casher, Chiral symmetry breaking in confining theories. Nucl. Phys. B **169**, 103 (1980)
30. Y. Nambu, G. Jona-Lasinio, Dynamical model of elementary particles based on an analogy with superconductivity. Phys. Rev. **122**, 345 (1961)
31. K.A. Intriligator, N. Seiberg, Lectures on supersymmetric gauge theories and electric-magnetic duality. Nucl. Phys. Proc. Suppl. **45BC**, 1 (1996). arXiv:hep-th/9509066.
32. J. Smit, *Introduction to Quantum Fields on a Lattice* (Cambridge University Press, Cambridge, 2002)
33. C. Gattringer, C.B. Lang, *Quantum Chromodynamics on the Lattice*. Lecture Notes in Physics, vol. 788 (Springer, Berlin, 2010)
34. R. Höllwieser, M. Faber, J. Greensite, U.M. Heller, Š. Olejník, Center vortices and the dirac spectrum. Phys. Rev. D **78**, 054508 (2008). arXiv:0805.1846 [hep-lat]
35. S. Adler, W.A. Bardeen, Absence of higher order corrections in the anomalous axial vector divergence equation. Phys. Rev. **182**, 1517 (1969)
36. A.A. Belavin, A.M. Polyakov, A.S. Shvarts, Yu.S. Tyupkin, Pseudoparticle solutions of the Yang-Mills equations. Phys. Lett. B **59**, 85 (1975)
37. E. Witten, Current algebra theorems for the U(1) Goldstone boson. Nucl. Phys. **B156**, 269 (1979)
38. G. Veneziano, U(1) without instantons. Nucl. Phys. **B159**, 213 (1979)
39. M. Engelhardt, Center vortex model for the infrared sector of Yang-Mills theory: topological susceptibility. Nucl. Phys. B **585**, 614 (2000). arXiv:hep-lat/0004013
40. H. Reinhardt, Topology of center vortices, in *Confinement, Topology, and Other Non-Perturbative Aspects of QCD*, ed. by J. Greensite, Š. Olejník (Kluwer Academic, Dordrecht, 2002), pp. 277–285. arXiv:hep-th/0204194
41. H. Reinhardt, O. Schröder, T. Tok, V. Zhukovsky, Quark zero modes in intersecting center vortex gauge fields. Phys. Rev. D **66**, 085004 (2002). arXiv:hep-th/0203012
42. R. Bertle, M. Engelhardt, M. Faber, Topological susceptibility of Yang-Mills center projection vortices. Phys. Rev. D **64**, 074504 (2001). arXiv:hep-lat/0104004
43. J. Greensite, S. Olejnik, Vortices, symmetry breaking, and temporary confinement in SU(2) gauge-Higgs theory. Phys. Rev. D **74**, 014502 (2006). arXiv:hep-lat/0603024
44. R. Bertle, M. Faber, J. Greensite, S. Olejnik, Center dominance in SU(2) gauge-Higgs theory. Phys. Rev. D **69**, 014007 (2004). arXiv:hep-lat/0310057

Abstract

Recent numerical tests of the center vortex mechanism.

The most extensive numerical tests to date of the center vortex mechanism have been carried out in recent years by Kamleh, Leinweber, and Trewartha (KLT) and Biddle [1–6] at the University of Adelaide. This chapter will be devoted to a brief exposition of their results.

7.1   Cooling and SU(3) Considerations

The KLT work is carried out in SU(3) gauge theory, and maximal center gauge fixing is then somewhat more involved than for the SU(2) case. Maximal center gauge is defined to be the gauge in which

$$R = \sum_x \sum_\mu |\mathrm{Tr} U_\mu(x)|^2 \qquad (7.1)$$

is maximized. As in the SU(2) case it is impossible in practice to find a global maximum, so instead a relaxation procedure is used to bring the link variables to a local maximum of R. The idea is again to sweep through the lattice site by site, and at each site to find a gauge transformation Ω with maximizes

$$R_x = \sum_\mu \left\{ |\mathrm{Tr}\Omega U_\mu(x)|^2 + |\mathrm{Tr} U_\mu(x - \hat{\mu})\Omega^\dagger|^2 \right\}. \qquad (7.2)$$

One then sets $U_\mu(x) \rightarrow \Omega U_\mu(x)$, $U_\mu(x - \hat{\mu})\Omega^\dagger$, and proceeds to the next site. But the question is how to find Ω. The SU(3) group is covered by three SU(2)

© Springer Nature Switzerland AG 2020   93
J. Greensite, *An Introduction to the Confinement Problem*, Lecture Notes
in Physics 972, https://doi.org/10.1007/978-3-030-51563-8_7

subgroups, embedded into SU(3) matrices as follows:

$$Q = \begin{pmatrix} q_{11} & q_{12} & 0 \\ q_{21} & q_{22} & 0 \\ 0 & 0 & 1 \end{pmatrix} \quad , \quad S = \begin{pmatrix} s_{11} & 0 & s_{12} \\ 0 & 1 & 0 \\ s_{21} & 0 & s_{22} \end{pmatrix} \quad , \quad T = \begin{pmatrix} 1 & 0 & 0 \\ 0 & t_{11} & t_{12} \\ 0 & t_{21} & t_{22} \end{pmatrix} . \tag{7.3}$$

Any SU(3) matrix can be obtained from a product QST, and the elements q_{ij}, s_{ij}, t_{ij} are each the elements of three 2×2 SU(2) matrices. Gauge fixing proceeds by choosing Ω to be one of these three SU(2) group elements embedded in SU(3), with SU(2) components

$$[\Omega]_{SU(2)} = g_4 \mathbb{1} - i \sum_{k=1}^{3} g_k \sigma_k \quad , \quad g_4^2 + \mathbf{g} \cdot \mathbf{g} = 1 . \tag{7.4}$$

One then finds

$$R_x = \sum_{i,j=1}^{4} \tfrac{1}{2} g_i a_{ij} g_j - \sum_{i=1}^{4} g_i b_i + c , \tag{7.5}$$

where the a_{ij}, b_i, c are elements of a symmetric 4×4 matrix, a 4-vector, and a constant respectively, which depend only on the matrix elements of the link variables $U_\mu(x), U_\mu(x - \hat{\mu})$. Explicit expressions can be found in [2]. The g_i which maximize R_x subject to the constraint in (7.4) can then be obtained by Lagrange multiplier techniques, and the link variables are transformed by the Ω which is so obtained. One then repeats the updating procedure with the other two SU(2) subgroups, and proceeds to the next lattice site.

Once the configuration is fixed to maximal center gauge, center projection is obtained by first computing, for each link,

$$\text{Tr} U_\mu(x) = r e^{i\theta} , \tag{7.6}$$

and then choosing $m \in \{-1, 0, 1\}$ so that $\frac{2\pi}{3} m$ is closest to θ. The center projected link is then

$$Z_\mu(x) = e^{i 2\pi m / 3} . \tag{7.7}$$

The vortex removed configurations are

$$\overline{U}_\mu(x) = Z_\mu^\dagger(x) U_\mu(x) . \tag{7.8}$$

Another standard technique, which is important to the KLT work, is known as "cooling." Lattice configurations in general contain not only the long range fluctuations relevant to non-perturbative phenomena such as confinement, topology, and chiral symmetry breaking, but also short range fluctuations associated with high momenta. These high momentum short range fluctuations are, from the non-perturbative point of view, a kind of noise which tends to conceal the interesting long range physics. A number of techniques have been introduced to smooth away these short wavelength fluctuations, the simplest of which is cooling. There are many variations, but the most direct goes as follows: Sweep through the lattice link by link, and at each link replace $U_\mu(x)$ by another SU(3) matrix element $U'_\mu(x)$ which minimizes the part of the action which contains that link, i.e. the object is to choose $U'_\mu(x)$ to maximize

$$S_{x,\mu} = \mathrm{ReTr}[U'_\mu(x)V_\mu(x)] \,, \tag{7.9}$$

where $V_\mu(x)$ is the sum of staples

$$V_\mu(x) = \sum_{\nu\neq\mu}(U_\nu(x+\hat{\mu})U_\mu^\dagger(x+\hat{\nu})U_\nu^\dagger(x)$$

$$+U_\nu^\dagger(x-\hat{\nu}+\hat{\mu})U_\mu^\dagger(x-\hat{\nu})U_\nu(x-\hat{\nu})\Big] . \tag{7.10}$$

If the group were SU(2), then V would be proportional to an SU(2) matrix, and the solution is simply

$$U'_\mu(x) = \frac{V_\mu^\dagger(x)}{\sqrt{\det V}} . \tag{7.11}$$

For SU(3) the solution is apply this algorithm to maximize $S_{x,\mu}$ using successively the SU(2) subgroup matrices Q, S, T, with the cooled link constructed from the product. Maximizing at each link constitutes one cooling sweep.

The cooling technique is not limited to the simple Wilson action, and in fact one may choose to cool using an improved action of some kind. Many of the results below were obtained using an "$O(a^4)$ improved" cooling technique, cf. [2] and references therein. There are other smoothing methods, and nowadays the gradient flow technique [7], which is a more controlled version of cooling, is usually the method of choice.

Cooling can be applied to the original "untouched" configurations, to the center projected "vortex only" configurations (7.7), and to the vortex removed configurations (7.8). As already mentioned, cooling or smoothing of some kind is important in computing topological properties of gauge configurations. The message of the KLT work, as we will see below, is that the untouched and vortex removed configurations converge after a number of cooling steps, in the sense that their topological properties, chiral symmetry breaking effects, string tensions, and even the mass spectrum extracted from the cooled configurations are very similar. The opposite is true for the vortex removed configurations, which almost entirely lack the non-perturbative phenomena that are characteristic of QCD.

7.2   Chiral Symmetry and Topology

Chiral symmetry breaking should manifest itself in the quark propagator, with an
effective mass much larger than the bare quark mass. As previously noted, on
the lattice there is a "doubling" problem, and in order to avoid such doubling
and still retain the chiral symmetry the preferred approach is to use the overlap
Dirac operator developed by Neuberger and Narayanan [8]. The quark propagator is
obtained from evaluating the expectation value of the inverse of the Dirac operator
in some definite gauge. In a covariant gauge on the lattice this propagator can be
shown to have the form (with color and Dirac indices implicit)

$$S(p) = \frac{Z(p)}{iq\!\!\!/ + M(p)}, \tag{7.12}$$

where q is a certain function of the lattice momentum p, cf. [9]. The quantity of
interest is the momentum-dependent mass function $M(p)$.

The quark mass function $M(p)$ for quarks with a 12 MeV bare mass, computed
in Landau gauge on untouched and vortex removed configurations, is displayed on
the left of Fig. 7.1. It can be seen that $M(p)$ is drastically reduced by vortex removal.
The figure on the right shows $M(p)$ computed on untouched and vortex-only
configurations after 10 cooling sweeps were applied to each. The mass functions
are almost identical, indicating that the vortex-only configurations have retained the
chiral symmetry breaking properties of the untouched configurations.

Another interesting observable is the instanton content in the untouched, vortex-
only, and cooled configurations. The strategy is to search the lattice for local maxima

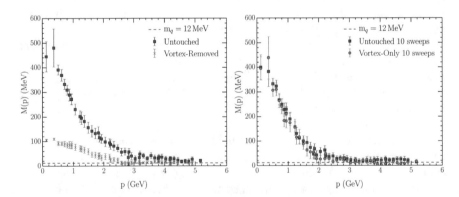

Fig. 7.1   The quark mass function $M(p)$, derived from the Landau gauge propagator corresponding to the overlap Dirac operator, at a bare mass of 12 MeV. Obviously vortex removal is a drastic
modification. By contrast, the left figure compares $M(p)$ for the untouched and vortex removed
configurations. The right figure is the comparison of $M(p)$ in the untouched and vortex-only
configurations, after each is subjected to 10 cooling sweeps, and it is clear that $M(p)$ in the two
types of configurations has converged. From Kamleh, Leinweber, and Trewartha [3]

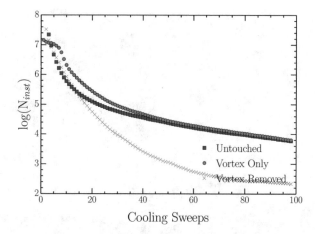

Fig. 7.2 A logarithmic plot of the number of instanton-like objects per configuration on a $20^3 \times 40$ lattice volume, comparing untouched, vortex-only and vortex-removed configurations versus cooling sweeps. From Trewartha et al. [2]

of the action density, and fit to

$$S_0 = \xi \frac{6}{\pi^2} \frac{\rho^4}{((x - x_0)^2 + \rho^2)^4)} \,. \tag{7.13}$$

Of course this method assumes a fairly smooth background, and for uncooled configurations is likely to result in many false positives just due to ultraviolet fluctuations. The results are shown in Fig. 7.2 (note the logarithmic scale). After a few sweeps of cooling to eliminate these false positives the untouched and vortex only instanton densities start to converge, while by comparison the instanton content of the vortex-removed configurations is down by about two orders of magnitude at the higher smoothing levels.

A closely related observable is the topological charge density $q(x)$, i.e. topological charge Q per unit volume, which in the continuum is essentially the integrand of Eq. (6.59). There are sophisticated methods for measuring this quantity on the lattice from the Dirac operator, see e.g. Chapter 7 of [10]. Visualizations of this observable are shown in Fig. 7.3. The left and right hand sides show the topological charge (positive in yellow, negative in blue) for configurations after 10 and 40 cooling sweeps, respectively, of a typical lattice configuration. The upper two plots are from an untouched configuration, the middle from the corresponding vortex-only configuration, and the lower two plots are the vortex-removed configuration. This is another indication of the similarity, at the non-perturbative level, of untouched and vortex-only configurations after moderate amounts of cooling, and the dissimilarity of the vortex-removed configuration.

It is worth noting that the positions of instantons, and the spatial distribution of topological charge, are not identical in the untouched configuration, and the

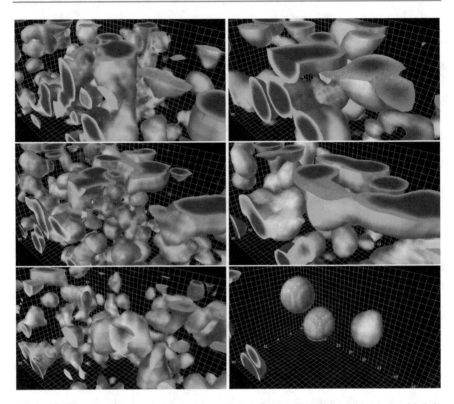

Fig. 7.3 The topological charge density on a typical lattice configuration, obtained after 10 (left) and (40) cooling sweeps. Positive topological charge is plotted in yellow, negative in blue. The upper plots are derived from the untouched configuration, the middle from the corresponding vortex-only configuration, and the bottom from the vortex-removed configuration. From Kamleh et al. [3]

vortex-only configuration obtained from center projection. It would appear that from the thin vortices of the center projected configuration, instantons and topological charge are generated after cooling with densities that are commensurate with the corresponding densities in the cooled but unprojected configuration. The evidence indicates that the existence of instantons in some way follows from the existence of thick center vortices.

We should finally note the extensive interactive graphics produced by Biddle et al. [6], which display the location of center vortices in $D = 3$ time slices of the lattice volume, in typical configurations generated in lattice Monte Carlo simulations. These visualisations highlight the frequent presence of writhing points in the vortex surface, and reveal an important role for vortex branching points in creating high topological charge density regimes in SU(3) gauge theory. Vortex branching in SU(3) gauge theory has also been studied by Spengler, Quandt, and Reinhardt in [11], who find that the vortex branching probability is lattice spacing independent.

7.3   The Hadron Mass Spectrum

The center projected string tension comes out a bit low in SU(3) gauge theory, only about 2/3 of the asymptotic string tension of the unprojected theory. But it is a curious fact that here again the string tensions of the untouched and vortex only configurations seem to converge after a few cooling sweeps, as seen in Fig. 7.4. The string tension vanishes for the vortex-removed configurations. So an interesting question is how the hadron spectrum differs in these three types of configurations. If chiral symmetry is truly restored in vortex-removed configurations, then this ought to be visible as a degeneracy in the spectrum. In particular one expects degeneracies between the π, a_0, the ρ, a_1, and N, Δ, where N is the nucleon. On the other hand, if the vortex-only configurations retain chiral symmetry breaking, then that would imply a light pion in the spectrum.

There is a standard technology for computing hadron masses [10]. One begins with a local operator on a timeslice t with the quantum numbers of the hadron of interest, and then averages over space in order to project to the zero momentum state. Denote this zero momentum operator $Q(t)$. Then lattice Monte Carlo simulations are used to compute the time correlation function

$$C(t) = \langle Q^{\dagger}(t)Q(0) \rangle = \sum_n \langle Q^{\dagger}(t)|n\rangle \langle n|Q(0)\rangle e^{-E_n t} . \tag{7.14}$$

The computation requires computing the expectation value of products of a quark and an antiquark propagator (for mesons) or three quark propagators (for baryons);

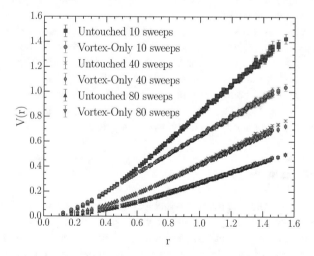

Fig. 7.4 The static quark potential on untouched (blue) and vortex-only (green) configurations after 10, 40, and 80 sweeps of cooling. Note that the 80 sweeps untouched potential is hidden behind the 80 sweeps vortex-only potential. From Trewartha et al. [2]

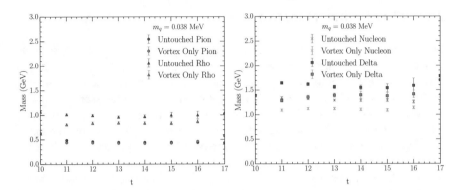

Fig. 7.5 A comparison of the effective masses $m_{eff}(t)$ of the low-lying mesons (left) and baryons (right), at bare quark mass $m_q = 38$ MeV, as computed on the untouched and vortex only configurations. (There appears to be a misprint on the figure, where m_q is quoted in MeV rather than GeV.) From Trewartha et al. [4]

the calculation is much simpler, if less realistic, if the action itself does not contain dynamical quarks. At sufficiently large t the correlation function will be dominated by the smallest mass state of the given quantum numbers, so we define the effective mass

$$m_{eff}(t) = \log \frac{C(t)}{C(t+1)} , \tag{7.15}$$

which should approximate the mass of the lowest mass hadron when $m_{eff}(t)$ is approximately t-independent.

The low-lying hadron spectrum computed from the untouched and vortex-only configurations in Ref. [4], on a $20^3 \times 40$ lattice volume after 10 cooling sweeps, is shown in Fig. 7.5. The untouched and vortex only results are qualitatively, and even to some extent quantitatively, in good agreement. By contrast, the hadron masses in the vortex removed configurations, shown in Fig. 7.6, are very different from the spectrum obtained from the untouched configurations, and display the Nucleon-Delta degeneracy associated with chiral symmetry restoration.

7.4   Conclusions

The results of the Adelaide group convincingly demonstrate that, after a few cooling steps, the non-perturbative properties of vortex-only configurations are virtually identical to those of the untouched configurations, after an equal number of cooling steps. These properties include the momentum-dependent quark mass function, the instanton and topological charge density, the string tension, and even the low-lying hadron mass spectrum. Vortex-removed configurations, as compared to the untouched configurations, are drastically different on all counts, and in particular

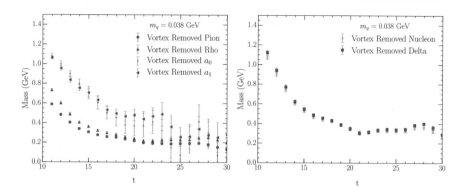

Fig. 7.6 The effective masses $m_{eff}(t)$ of the low-lying mesons (left) and baryons (right), at bare quark mass $m_q = 38$ MeV, as computed on the vortex-removed configurations. From Trewartha et al. [4]

appear to have a chiral symmetric mass spectrum. These results are perhaps the strongest evidence to date that the non-perturbative properties of QCD are deeply entwined with the dominance of center vortex configurations in the QCD vacuum.

References

1. D. Trewartha, W. Kamleh, D. Leinweber, Evidence that centre vortices underpin dynamical chiral symmetry breaking in SU(3) gauge theory. Phys. Lett. B **747**, 373 (2015). arXiv:1502.06753 [hep-lat]
2. D. Trewartha, W. Kamleh, D. Leinweber, Connection between center vortices and instantons through gauge-field smoothing. Phys. Rev. D **92**(7), 074507 (2015). arXiv:1509.05518 [hep-lat]
3. W. Kamleh, D.B. Leinweber, D. Trewartha, Centre vortices are the seeds of dynamical chiral symmetry breaking. PoS LATTICE **2016**, 353 (2017). arXiv:1701.03241 [hep-lat]
4. D. Trewartha, W. Kamleh, D. Leinweber, Centre vortex removal restores chiral symmetry. J. Phys. G **44**(12), 125002 (2017). arXiv:1708.06789 [hep-lat]
5. J.C. Biddle, W. Kamleh, D.B. Leinweber, Gluon propagator on a center-vortex background. Phys. Rev. D **98**(9), 094504 (2018). arXiv:1806.04305 [hep-lat]
6. J.C. Biddle, W. Kamleh, D.B. Leinweber, Visualisation of Centre Vortex Structure. arXiv:1912.09531 [hep-lat]
7. M. Lüscher, Properties and uses of the Wilson flow in lattice QCD. J. High Energy Phys. **1008**, 071 (2010). arXiv:1006.4518 [hep-lat]
8. R. Narayanan, H. Neuberger, A construction of lattice chiral gauge theories. Nucl. Phys. B **443**, 305 (1995). hep-th/9411108
9. F.D.R. Bonnet et al., [CSSM Lattice Collaboration], Overlap quark propagator in Landau gauge. Phys. Rev. D **65**, 114503 (2002) [hep-lat/0202003]
10. C. Gattringer, C.B. Lang, *Quantum Chromodynamics on the Lattice*. Lecture Notes on Physics, vol. 788 (Springer, Berlin, 2010)
11. F. Spengler, M. Quandt, H. Reinhardt, Branching of center vortices in SU(3) lattice gauge theory. Phys. Rev. D **98**(9), 094508 (2018). https://doi.org/10.1103/PhysRevD.98.094508. arXiv:1810.04072 [hep-th]

Confinement from Center Vortices III

8

Abstract

Casimir scaling and the center vortex picture. The random surface model. Center vortices as classical solutions of an effective gauge theory.

There are still a number of important questions relating to center vortices and confinement which must be addressed. First in importance is the issue of Casimir scaling. We have seen that there is some intermediate range of distances for which the string tension is proportional to the quadratic Casimir of the quark color representation; the dependence shifts to N-ality only at asymptotic distances. How is Casimir scaling to be understood, in the context of the center vortex mechanism? A second question is whether vortices can be understood as the classical solutions of some effective gauge theory action, which is relevant to large scales. Finally, is it possible to devise a simple theory of fluctuating vortex surfaces, which would describe the behavior of non-abelian gauge theories in the infrared regime?

8.1   Casimir Scaling and Vortex Thickness

Let's begin with Casimir scaling. Although the asymptotic string tension only depends on N-ality, so that for $SU(2)$

$$\sigma_j = \begin{cases} \sigma_{1/2} & j = \text{half-integer} \\ 0 & j = \text{integer} \end{cases} , \qquad (8.1)$$

there is still an intermediate range of distances where Casimir scaling applies (at least approximately), i.e. for $SU(2)$

$$\sigma_j \propto j(j+1) . \qquad (8.2)$$

© Springer Nature Switzerland AG 2020
J. Greensite, *An Introduction to the Confinement Problem*, Lecture Notes in Physics 972, https://doi.org/10.1007/978-3-030-51563-8_8

How do vortices fit in, since they are motivated by (and seem to only give rise to) N-ality dependence?

The essential observation is that while j = integer Wilson loops are unaffected by *thin* center vortices, they can be affected by *thick* center vortices, if the vortex core overlaps the loop. But how thick are center vortices? We have already seen that, in $D = 3$ dimensions, the observed adjoint string breaking occurs at 1.25 fm [1], when "physical units" are obtained by fixing the string tension at $\sigma = (440\,\text{MeV})^2$. It was also found, in $D = 4$ dimensions, that the center vortex free energy drops to zero at around 1.25 fm [2]. Finally, from Fig. 6.3 of Chap. 6, one may estimate that $W_1/W_0 \approx -1$ for 5×5 vortex-limited Wilson loops, which translates, at $\beta = 2.3$ into an extension of roughly 0.85 fm. Each of these observations suggests a vortex thickness of around one fermi. This is not so small, and the question is whether this finite thickness makes a difference for Wilson loops of dimensions less than one fermi.

A simple model, which shows how Casimir scaling at intermediate distances might emerge due to the finite thickness of vortices, was put forward in [3], improving on an earlier suggestion in [4].[1] Consider the projection of a vortex onto the plane of a planar Wilson loop; we will refer to this projection as a vortex "domain" in the plane. In this model, it is assumed that the effect of a domain (2D cross-section of a vortex) on a planar Wilson loop holonomy is to multiply the holonomy by a group element

$$G(\alpha^n, S) = S \exp[i\alpha^n \cdot \mathbf{H}]S^\dagger , \qquad (8.3)$$

where the $\{H_i\}$ are generators of the Cartan subalgebra (the largest subset of group generators which commute with one another), S is a random group element, $\{\alpha_i^n\}$ depends on the location of the domain relative to the loop, and n indicates the domain type. If the domain lies entirely within the planar area enclosed by the loop, then

$$\exp[i\alpha^n \cdot \mathbf{H}] = z_n \mathbb{1}_N , \qquad (8.4)$$

where

$$z_n = e^{2\pi in/N} \in Z_N \qquad (8.5)$$

and $\mathbb{1}_N$ is the $N \times N$ unit matrix. At the other extreme, if the domain is entirely outside the planar area enclosed by the loop, then

$$\exp[i\alpha^n \cdot \mathbf{H}] = \mathbb{1}_N . \qquad (8.6)$$

---

[1] See also Deldar [5]. A still earlier idea along these lines was put forward by Cornwall [6].

For a Wilson loop in representation r, the average contribution from a domain will be

$$\mathscr{G}_r[\alpha^n]\mathbb{1}_{d_r} = \int dS\, S \exp[i\alpha^n \cdot \mathbf{H}]S^\dagger$$

$$= \frac{1}{d_r}\chi_r\Big[\exp[i\alpha^n \cdot \mathbf{H}]\Big]\mathbb{1}_{d_r}, \qquad (8.7)$$

where d_r is the dimension of representation r, $\mathbb{1}_{d_r}$ is the $d_r \times d_r$ unit matrix, and χ_r is the group character in representation r.

Consider, e.g., $SU(2)$ lattice gauge theory, choosing $H = L_3$. The center subgroup is Z_2, and there are two types of domains, corresponding to $z_0 = 1$ and $z_1 = -1$. Let f_1 represent the probability that the midpoint of a z_1 domain is located at any given plaquette in the plane of the loop, with f_0 the corresponding probability for a $z_0 = 1$ domain. Let us also assume that the probabilities of finding domains of either type centered at any two plaquettes in a plane are independent. Then, for a planar loop C,

$$W_j(C) \approx \prod_x \Big\{(1 - f_1 - f_0) + f_1\mathscr{G}_j[\alpha_C^1(x)] + f_0\mathscr{G}_j[\alpha_C^0(x)]\Big\}W_j^{pert}(C)$$

$$= \exp\Big[\sum_x \log\Big\{(1 - f_1 - f_0) + f_1\mathscr{G}_j[\alpha_C^1(x)]$$

$$+ f_0\mathscr{G}_j[\alpha_C^0(x)]\Big\}\Big]W_j^{pert}(C), \qquad (8.8)$$

where the product and sum over x runs over all plaquette positions in the plane of the loop C, and $\alpha_C^n(x)$ depends on the position of the vortex midpoint x relative to the location of loop C. The expression $W_j^{pert}(C)$ contains the short-distance, perturbative contribution to $W_j(C)$; this will just have a perimeter-law falloff.

The question is what to use as an ansatz for $\alpha_C^n(x)$. The ansatz of Ref. [3] was motivated by the idea that the magnetic flux in the interior of vortex domains fluctuates almost independently, in subregions of extension l, apart from the restriction that the total flux results in a center element. For the $SU(2)$ gauge group, this leads to

$$\left(\alpha_C^1(x)\right)^2 = \frac{A_v}{2\mu}\left[\frac{A}{A_v} - \frac{A^2}{A_v^2}\right] + \left(2\pi\frac{A}{A_v}\right)^2$$

$$\left(\alpha_C^0(x)\right)^2 = \frac{A_v'}{2\mu}\left[\frac{A}{A_v'} - \frac{A^2}{A_v'^2}\right], \qquad (8.9)$$

where A_v, A_v' are the cross-sectional areas of the $n = 1$ and $n = 0$ domains, respectively, and A is the area of the domain which is contained within the interior of the minimal area of the loop. The parameter μ controls the magnitude of the

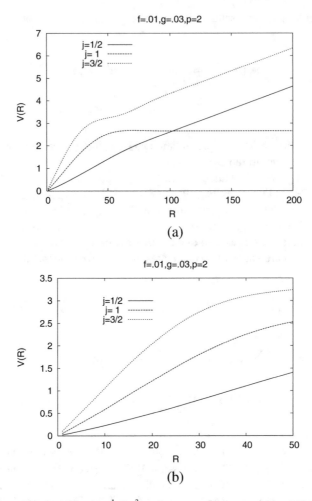

Fig. 8.1 The static potential for $j = \frac{1}{2}, 1, \frac{3}{2}$ static sources, for vortex width = 100, in the distance range: (**a**) $R \in [0, 200]$; and (**b**) $R \in [0, 50]$. From Ref. [3]

magnetic fluctuations in the subregions. The result is a vortex-induced potential which is proportional to the quadratic Casimir of the representation at small R (compared to the vortex thickness), and depends only on the N-ality at large R. A sample calculation for the $j = \frac{1}{2}, 1, \frac{3}{2}$ representations is shown in Fig. 8.1. Details of the calculation are found in Ref. [3].

Casimir scaling for the smaller loops is not an accident, and not specific to the $SU(2)$ gauge group. In generalizing to $SU(N)$ we must take into account the fact that there are N types of center domains, so that

$$\langle W_r(C) \rangle$$

$$\approx \prod_x \left\{ 1 - \sum_{n=0}^{N-1} f_n \left(1 - \text{Re}\mathscr{G}_r[\alpha_C^n(x)] \right) \right\} W_r^{pert}(C)$$

$$= \exp \left[\sum_x \log \left\{ 1 - \sum_{n=0}^{N-1} f_n \left(1 - \text{Re}\mathscr{G}_r[\alpha_C^n(x)] \right) \right\} \right] W_r^{pert}(C) , \qquad (8.10)$$

where

$$f_n = f_{N-n} \quad , \quad \mathscr{G}_r[\alpha^n] = \mathscr{G}_r^*[\alpha^{N-n}] . \qquad (8.11)$$

For very small loops, which are associated with small α, we have that

$$1 - \mathscr{G}_r[\alpha^n] \approx \tfrac{1}{2} \alpha_i^n \alpha_j^n \frac{1}{d_r} \text{Tr}[H^i H^j]$$

$$= \frac{1}{2(N^2 - 1)} \alpha^n \cdot \alpha^n C_r , \qquad (8.12)$$

where C_r is the quadratic Casimir for representation r. Once again, if $\alpha^n \cdot \alpha^n$ is proportional to the area of the vortex in the interior of the loop, for small loops, then the result for small loops is a linear potential whose string tension is proportional to the quadratic Casimir. For large loops, if the vortex domain lies entirely in the loop interior, we must have

$$\frac{1}{d_r} \chi_r \left[\exp[i\alpha^n \cdot \mathbf{H}] \right] = e^{ikn\pi/N} , \qquad (8.13)$$

where k is the N-ality of representation r. Again we obtain a linear potential, whose string tension depends only on the N-ality. It is possible to choose the f_n so that for very large loops the k-string tensions follow either the Sine Law, or are proportional to the Casimir $C_{r_{min}}$ of the lowest dimensional representation of N-ality k; cf. Ref. [7] for a full discussion of this point. Recently Junior, Oxman, and Simoes have shown [8], in the context of a $D = 2 + 1$ dimensional model of center vortices, that the asymptotic k-string tensions obey Casimir scaling.

G_2 gauge theory is a special case of particular interest, because the center of the G_2 group consists only of the unit element. In this theory there would be only one type of vortex domain, corresponding to the single element of the trivial center group. One would then expect all color representations to give rise to Casimir scaling in the intermediate distance regime, and flatten out asymptotically. Casimir scaling in this theory has in fact been observed, cf. Ref. [9].

8.2   What About "Gluon Confinement"?

Center vortices account for the asymptotic string tension of the static quark potential. This asymptotic string tension depends only on the N-ality, not the dimension,
of the color representation, and the string tension of N-ality = 0 representations is
zero, as it should be. But then, one may ask, what about gluons? These are in the
adjoint representation, so their N-ality is zero, yet gluons are not members of the
asymptotic particle spectrum. So what accounts for the "confinement" of gluons?

To understand this point, it is worth reflecting on the fact that one would
never encounter superheavy nuclei, of some very large but fixed baryon number
A, with charge greater than or equal to some critical amount $Q_c \ll A$, even if
such nuclei were stable to nuclear decay [10]. The reason is simple: the vacuum
of quantum electrodynamics at any instant is a superposition of states containing
virtual e^+e^- pairs, as seen in Fig. 8.2. If a superheavy nucleus-antinucleus pair
were somehow created with electric charges exceeding the critical value (Fig. 8.3),
then the associated electric field energy would be so strong that it is energetically
favorable for some of the virtual electrons and positrons to bind to these objects,
screening their charges down below the critical value (Fig. 8.4). Nuclear charge at
Q_c and above is therefore "confined" in nuclei of fixed baryon number A, in the
sense that nuclei with such high electric charges will never be observed.

Similar reasoning (illustrated in Figs. 8.5, 8.6, 8.7) explains why isolated
particles such as gluons, with N-ality = 0 color charge, are absent in the asymptotic
spectrum. The QCD vacuum contains, in addition to virtual quarks and antiquarks,
also virtual gluons. As two particles in the adjoint representation become widely

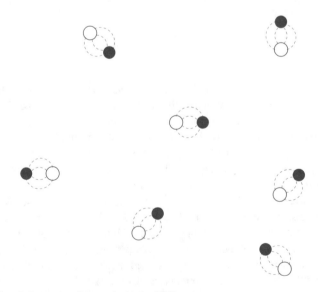

Fig. 8.2 Virtual electron-positron pairs in the QED vacuum

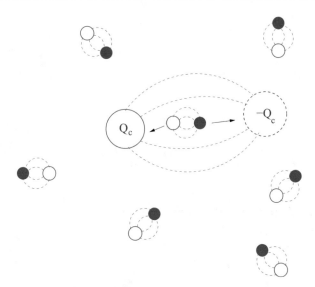

Fig. 8.3 Introduction of nuclei with charges $\pm Q_c$ beyond the threshold value

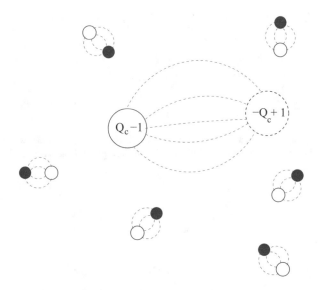

Fig. 8.4 Response of the vacuum is to screen the nuclear charge below the critical value

separated and enter the Casimir scaling regime, the energy of the color electric field becomes large. At some point it becomes energetically favorable for virtual gluons to bind to the sources, resulting in two color-singlet states. This is essentially the string-breaking mechanism discussed previously. While it is only a matter of semantics whether one calls this process confinement, I prefer, for reasons discussed

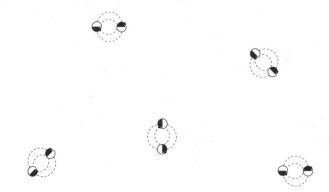

Fig. 8.5 Virtual gluons in the QCD vacuum

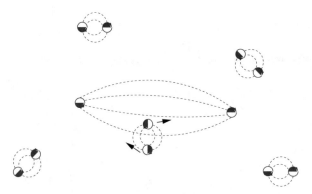

Fig. 8.6 Two gluons (or other particles of adjoint color charge) become widely separated, momentarily. The effective coupling at this separation is large, and the associated color electric field energy is large

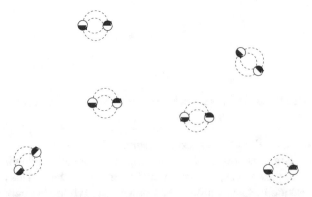

Fig. 8.7 Response of the nearby gluons is to reduce the color charge separation via the mechanism of string breaking

in Chap. 3, to refer to it as color screening, and reserve the term "confinement" for magnetic disorder.

These remarks about gluon confinement also apply to G_2 gauge theory. As already noted, the center of G_2 is trivial, and this has led some authors [11] to argue that since G_2 is a confining gauge theory with a trivial center, center symmetry (and the vortex mechanism) cannot be necessary for confinement. However, since all of the usual order parameters for confinement, such as Wilson loops and Polyakov loops, exhibit a non-confining behavior in the G_2 case, it brings us back to question of what is meant by the term "confinement" (a question we will return to in Chap. 15). If all that is meant is that the theory has a color-singlet spectrum, then it is certainly true that vortices are not required. As we have seen, a color-singlet spectrum is also obtained in a gauge-Higgs theory. If, on the other hand, by confinement we mean a certain property of the vacuum state, i.e. magnetic disorder at arbitrarily large scales, then G_2 gauge theory does not have this property. From this point of view, the G_2 gauge group is better regarded an example of, rather than a counter-example to, the relevance of vortices to confinement [3]. No center vortices means no asymptotic string tension. The absence of color non-singlet quarks in G_2 gauge theory is no different, in principle, from the absence of colored gluons in $SU(N)$ theory. In both cases, the absence of colored states can be attributed to a color screening mechanism.

8.3   The Random Surface Model

We have seen that center projected field configurations can account for most of the long-range physics associated with Yang-Mills theories; i.e. confinement, chiral symmetry breaking, topological susceptibility, and spatial string tension in the deconfinement phase. This raises a natural question: Can one construct a field theory of vortices alone, which would capture the same physics? Such a theory was introduced by Engelhardt and Reinhardt in [12], and further developed by Engelhardt and collaborators [13–17].

Vortices are surfaces in $D = 4$ dimensions, so the Engelhardt-Reinhardt model is a theory of fluctuating random surfaces; i.e. a kind of string theory. In its simplest form, the model contains only one dimensionless coupling c, and a fixed lattice spacing a. The finite lattice spacing corresponds to the fact that center vortices in Yang-Mills configurations have a finite thickness; it therefore makes no sense to take the $a \to 0$ limit, which would describe infinitely thin vortices. The vortex sheets, in the $SU(2)$ gauge theory, are closed surfaces on a hypercubic lattice, and the lattice action is simply

$$S_V = cN_{bend} , \tag{8.14}$$

where N_{bend} is the number of cases in which two neighboring vortex plaquettes share a link, but lie in orthogonal planes. In other words, the action density is proportional to the extrinsic curvature of the vortex worldsheet. This is in contrast to a Nambu-style string action, in which the action is proportional to the area of the world sheet. In fact one could add such a term, with an associated coupling s, but Engelhardt and Reinhardt have shown that there is a line of constant physics in the c–s coupling plane, so one may as well drop the Nambu term and just work with the extrinsic curvature.

In the $SU(2)$ version of the random surface model, one evaluates Wilson loops in the obvious way, assigning a value $(-1)^n$ to a Wilson loop, where n is the number of vortex piercings of the minimal area of the loop, and averaging this quantity over an ensemble of vortex configurations generated, via Monte Carlo simulation, with the action (8.14). One can vary temperature by varying the extension of the lattice in the time direction, and, in the $SU(2)$ case, a second-order phase transition is found. Then the coupling c is fixed to reproduce the dimensionless quantity $T_C/\sqrt{\sigma}$, where T_C is the deconfinement transition temperature, and σ is the string tension, with both quantities obtained in the random surface model in lattice units. Given c, the lattice spacing a is fixed by setting $\sigma/a^2 = (440\,\text{MeV})^2$.

With coupling c and lattice spacing a in hand, the random surface model becomes predictive, and the following results have been obtained for the $SU(2)$ version of the theory:

- The spatial string tension in the deconfined phase vs. temperature [12]. The values are in good agreement with the full theory.
- The topological susceptibility [13] (Fig. 8.8). Again the values are in good agreement with the full theory, e.g. the topological susceptibility χ in the full theory, at zero temperature, is known to be $\chi^{1/4} \approx 190\,\text{MeV}$.
- The chiral condensate [14].

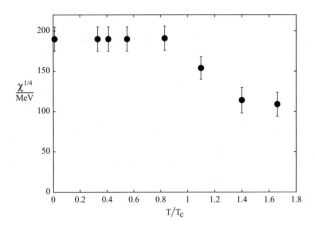

Fig. 8.8 Fourth root of the topological susceptibility vs. temperature, in the random surface model. From Engelhardt [13]

Fig. 8.9 The electric flux tube between three static quarks in an $SU(3)$ gauge theory can have either the "Δ" (left) or the "Y" (right) configuration shown. The random surface model predicts a "Y" configuration, in agreement with the known result [18, 19] found via standard Monte Carlo simulations

It was already mentioned that for the $SU(2)$ theory, the model finds a second order phase transition, in agreement with the known result for the full theory. The random surface model can be extended to the $SU(3)$ [15] and larger gauge groups, and for $SU(3)$ it is found that the deconfinement transition is weakly first-order, and, for baryons, the area law behaves as though the electric flux tubes between quarks are in a "Y" configuration, rather than a Δ configuration, as shown in Fig. 8.9. These features are also in agreement with the known properties of $SU(3)$ lattice gauge theory.

In order to measure the topological susceptibility, it is necessary to measure the topological charge density. As already noted in the previous chapter, a non-zero topological charge density is associated with vortex plaquettes which meet at a point, and which are orthogonal to one another; these are the "writhing" and self-intersection points displayed in Fig. 6.16. However, in the Z_2 random surface model there is no indication of the sign of $F_{\mu\nu}$, which is needed to define the topological charge. This is handled by the surprising procedure of assigning orientations (signs) stochastically to vortex plaquettes; this amounts to inserting monopole lines on the vortex surfaces with some weighting. One would think that this would introduce some new coupling parameter, and that while one might be able to fit a known value of χ, there would be no predictive power. It turns out however, for geometrical reasons explained in [13], that χ is insensitive to the density of monopole lines, and therefore the measurement of this quantity is a genuine prediction of the model, whose result is displayed in Fig. 8.8.

As also noted in the last section, the chiral condensate $\langle \overline{\psi}\psi \rangle$ is proportional, according to the Banks-Casher relation, to the eigenvalue density $\rho(0)$ of the zero modes of the Dirac operator. This has also been computed in the random surface model, by assigning a continuum gauge-field configuration to a vortex surface, and evaluating the continuum Dirac operator in this configuration. The results derived from the model are displayed in Fig. 8.10, and we see that the model predicts $\rho(0) > 0$ at temperatures below the deconfinement transition, and therefore a non-vanishing chiral condensate and chiral symmetry breaking, while $\rho(0) = 0$ above the transition temperature, which means that chiral symmetry is restored.

In general, the random surface model works remarkably well for $SU(2)$ and $SU(3)$ gauge theories, with only a single dimensionless coupling and lattice spacing required to account for a wide range of non-perturbative phenomena. For larger gauge groups, where there is more than one type of vortex and various vortex

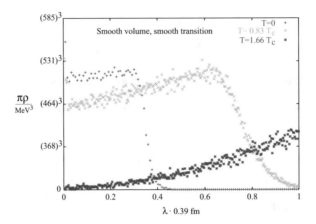

Fig. 8.10 Dirac spectral density in the random vortex model. From Engelhardt [14]

joinings are possible, it seems to be necessary to introduce additional couplings to describe the data, and consequently the model is much less predictive in those cases [17].

8.4   Vortices as Solitons

The only classical solutions which are known in pure gauge theories are the instanton solutions discovered by Belavin et al. [20], and their finite temperature variants known as *calorons*, to be discussed in the next chapter. If matter fields are added, there are other interesting possibilities. In particular, in an $SU(N)$ gauge theory, suppose we add enough matter fields in the adjoint color representation to give a mass, via the Higgs mechanism, to all of the gauge bosons. In that case, the theory typically contains center vortices as classical solutions.

A particularly interesting case has long been advocated by Cornwall [21] on the grounds that the dynamics of a pure gauge field should generate a mass for the gluon, and that this effect could be represented by an effective local action with a gauge-invariant mass term. Consider the action

$$S = \int d^4 x \left\{ \frac{1}{2g^2} \mathrm{Tr}[F_{\mu\nu}^2] + \frac{m^2}{g^2} \mathrm{Tr}[(D_\mu U)U^{-1}]^2 \right\} , \qquad (8.15)$$

where $U(x)$ is an $N \times N$ unitary-matrix valued field, transforming under an $SU(N)$ gauge transformation G as $U(x) \rightarrow G(x)U(x)$. This theory has center vortex solutions [21, 22], specified by a closed two-surface S, and a gauge field given by the integral over this surface:

$$A_\mu(x) = -2\pi i \, Q_a \varepsilon_{\mu\nu\alpha\beta} \partial_\nu \int d\sigma_{\alpha\beta} \, \tfrac{1}{2}[\Delta_m(x - x'(\sigma)) - \Delta_0(x - x'(\sigma))] , \qquad (8.16)$$

where $\Delta_{m,0}(x)$ is the standard propagator for a free scalar of mass m or mass zero, respectively, and Q_a is any one of the diagonal matrices

$$Q_a = \text{diag}\left[\frac{1}{N}, \frac{1}{N}, \ldots, -1 + \frac{1}{N}, \frac{1}{N}, \ldots\right] \tag{8.17}$$

with the $-1 + 1/N$ entry in the a-th position. These matrices have the property that

$$e^{2\pi i Q_a} = e^{2\pi i/N} . \tag{8.18}$$

In the special case that the 2-surface is the $z - t$ plane (so at a fixed time the middle of the center vortex runs along the z-axis), the solution (in cylindrical coordinates z, ρ, ϕ) is

$$\mathbf{A}_\mu(\mathbf{x}) = -i\widehat{\phi}Q_a\left[\frac{1}{\rho} - mK_1(m\rho)\right] , \tag{8.19}$$

where K_1 is a modified Bessel function. It is not hard to see that for large Wilson loops (at large ρ, in this example) which are linked topologically to the surface S, the loop holonomy $P \exp[i \oint dx_\mu A_\mu]$ approximates the center element $\exp[2\pi i/N]$. Center vortices carrying larger amounts of center flux are constructed by replacing a single Q_a with a sum of different Q_a's with different values of a.

The effective theory specified by (8.15) is quite rich. In addition to vortex solutions, the theory also contains "nexus" solutions [22] which correspond to magnetic monopoles running along the vortex sheets. These have, in fact, been seen in numerical simulations (as will be discussed in the next chapter) and are required for the generation of topological charge, as realized independently by Engelhardt and Reinhardt [13, 23]. In the case of SU(3), the effective theory predicts a "Y" configuration for baryonic (i.e. three-quark) flux-tube states [24], in agreement with numerical simulations [18].

In an alternative approach, a one loop effective action in a center vortex background was computed by Diakonov and Maul [25], and it was found that the effective action is minimized for vortices of a finite transverse thickness. Based on the fact that the effective action is negative at the minimum, these authors have argued that the perturbative vacuum may be unstable with respect to vortex creation.

8.5 Critique

I have presented a fair amount of evidence in favor of the center vortex scenario of confinement. There are still, however, some unsettled issues.

First of all, in numerical simulations, there is the question of what to do about Gribov copies in maximal center gauge. Should these be averaged over, or should one attempt to fix to a "best" copy? Apart from a question of principle, the strategy chosen makes a substantial quantitative difference (perhaps 25%) to the projected string tension, with the "average copy" procedure yielding the better results. Secondly, the vortex scenario gives us no real insight into the origin of the Lüscher term in the static quark potential. Presumably this term is hidden in some subtle correlations among vortex piercings, but we can't claim that it appears naturally. Finally, I have mainly shown numerical results for the $SU(2)$ gauge group. The corresponding numerical results in $SU(3)$ have both good and less good features, from the point of view of the vortex scenario. In $SU(3)$ the data for vortex-limited Wilson loops does trend to the desired result [26]

$$\frac{W_1(C)}{W_0(C)} \to \exp[\pm 2\pi i/3] \tag{8.20}$$

for large loops, and vortex removal sends the string tension to zero. Both of these findings support the vortex picture. On the other hand, in $SU(3)$ the center-projected string tension comes out to only about $\frac{2}{3}$ the full string tension [26], which is substantially worse than for the $SU(2)$ case. However, there is evidence that the string tensions of the full and vortex-only configurations in SU(3) gauge theory converge after a few cooling steps, as discussed in the previous chapter.

References

1. P. de Forcrand, S. Kratochvila, Observing string breaking with Wilson loops. Nucl. Phys. Proc. Suppl. **119**, 670 (2003). arXiv: hep-lat/0209094
2. T. Kovács, E. Tomboulis, Computation of the vortex free energy in SU(2) gauge theory. Phys. Rev. Lett. **85**, 704 (2000). arXiv:hep-lat/0002004
3. J. Greensite, K. Langfeld, Š. Olejník, H. Reinhardt, T. Tok, Color screening, Casimir scaling, and domain structure in G(2) and SU(N) gauge theories. Phys. Rev. D **75**, 034501 (2007). arXiv:hep-lat/0609050
4. M. Faber, J. Greensite, S. Olejnik, Casimir scaling from center vortices: Towards an understanding of the adjoint string tension. Phys. Rev. D **57**, 2603 (1998). arXiv:hep-lat/9710039
5. S. Deldar, Potentials between static SU(3) sources in the fat-center-vortices model. JHEP **0101**, 013 (2001). arXiv:hep-ph/9912428
6. J.M. Cornwall, in *Proceedings of the Workshop on Non-Perturbative Quantum Chromodynamics*, ed. by K.A. Milton, M.A. Samuel (Birkhäuser, Stuttgart, 1983)
7. J. Greensite, S. Olejnik, K-string tensions and center vortices at large N. JHEP **0209**, 039 (2002). arXiv:hep-lat/0209088
8. D.R. Junior, L.E. Oxman, G.M. Simões, 3D Yang-Mills confining properties from a non-Abelian ensemble perspective. JHEP **2001**, 180 (2020). https://doi.org/10.1007/JHEP01(2020)180. arXiv:1911.10144 [hep-th]
9. L. Liptak, S. Olejnik, Casimir scaling in G(2) lattice gauge theory. Phys. Rev. D **78**, 074501 (2008). arXiv:0807.1390 [hep-lat]
10. J. Madsen, Universal charge-radius relation for subatomic and astrophysical compact objects. Phys. Rev. Lett. **100**, 151102 (2008)
11. K. Holland, P. Minkowski, M. Pepe, U.J. Wiese, Exceptional confinement in G(2) gauge theory. Nucl. Phys. B **668**, 207 (2003). arXiv:hep-lat/0302023
12. M. Engelhardt, H. Reinhardt, Center vortex model for the infrared sector of Yang-Mills theory: Confinement and deconfinement. Nucl. Phys. B **585**, 591 (2000). arXiv:hep-lat/9912003
13. M. Engelhardt, Center vortex model for the infrared sector of Yang-Mills theory: Topological susceptibility. Nucl. Phys. B **585**, 614 (2000). arXiv:hep-lat/0004013

14. M. Engelhardt, Center vortex model for the infrared sector of Yang-Mills theory: Quenched Dirac spectrum and chiral condensate. Nucl. Phys. B **638**, 81 (2002). arXiv:hep-lat/0204002
15. M. Engelhardt, Center vortex model for the infrared sector of SU(3) Yang-Mills theory: Baryonic potential. Phys. Rev. D **70**, 074004 (2004). arXiv:hep-lat/0406022
16. M. Quandt, H. Reinhardt, M. Engelhardt, Center vortex model for the infrared sector of SU(3) Yang-Mills theory: Vortex free energy. Phys. Rev. D **71**, 054026 (2005). arXiv:hep-lat/0412033
17. M. Engelhardt, B. Sperisen, Confinement and center vortex dynamics in different gauge groups. AIP Conf. Proc. **892**, 176 (2007). arXiv:hep-lat/0610122
18. T.T. Takahashi, H. Suganuma, Detailed analysis of the gluonic excitation in the three-quark system in lattice QCD. Phys. Rev. D **70**, 074506 (2004). arXiv:hep-lat/0409105
19. C. Alexandrou, P. de Forcrand, O. Jahn, The ground state of three quarks. Nucl. Phys. Proc. Suppl. **119**, 667 (2003). arXiv:hep-lat/0209062
20. A.A. Belavin, A.M. Polyakov, A.S. Shvarts, Yu.S. Tyupkin, Pseudoparticle solutions of the Yang-Mills equations. Phys. Lett. B **59**, 85 (1975)
21. J. Cornwall, Quark confinement and vortices in massive gauge invariant QCD. Nucl. Phys. B **157**, 392 (1979)
22. J.M. Cornwall, On topological charge carried by nexuses and center vortices. Phys. Rev. D **65**, 085045 (2002). arXiv:hep-th/0112230
23. M. Engelhardt, H. Reinhardt, Center projection vortices in continuum Yang-Mills theory. Nucl. Phys. B **567**, 249 (2000). arXiv:hep-th/9907139
24. J.M. Cornwall, On the center-vortex baryonic area law. Phys. Rev. D **69**, 065013 (2004). arXiv:hep-th/0305101
25. D. Diakonov, M. Maul, Center-vortex solutions of the Yang-Mills effective action in three and four dimensions. Phys. Rev. D **66**, 096004 (2002). arXiv:hep-lat/0204012
26. K. Langfeld, Vortex structures in pure SU(3) lattice gauge theory. Phys. Rev. D **69**, 014503 (2004). arXiv:hep-lat/0307030

Monopoles, Calorons, and Dual Superconductivity

9

Abstract

Confinement via magnetic monopoles; confinement in compact QED$_3$. Electric-magnetic duality, and the dual superconductor scenario for confinement. Abelian projection as a monopole locator in pure gauge theories, vortices as a monopole-antimonopole chain. The caloron picture.

In a type II superconductor, magnetic fields are collimated into flux tubes known as *Abrikosov vortices*. If stable magnetic monopoles existed, and could be pair produced within a type II superconductor, then a magnetic flux tube would run between widely separated monopoles and antimonopoles, and the static monopole potential would rise linearly with monopole separation.[1] It was suggested by 't Hooft [2] and Mandelstam [3] in the mid 1970s that the QCD vacuum might resemble a type II superconductor, except with a reversal of the roles of the electric and magnetic fields. Instead of a condensate of electrically charged particles (the Cooper pairs of a superconductor), there would be a condensate of magnetic charge; and magnetic monopoles were invoked to fill this role. Then, instead of collimation of the color magnetic field, it would be the color electric field that would be squeezed into a flux tube, and color electric, rather than magnetic, charges would be confined by a linear potential.

This idea of "dual" superconductivity is simple and attractive, and has had an enormous influence on people's thinking about confinement, but it is essentially an abelian mechanism, because the magnetic monopoles of interest are normally defined with respect to abelian groups. In the context of a non-abelian gauge theory, it is necessary to single out an abelian subgroup, and this is done with the help of

---

[1]For a review of magnetic monopole solutions in abelian and non-abelian gauge theories, cf. Coleman [1].

© Springer Nature Switzerland AG 2020

J. Greensite, *An Introduction to the Confinement Problem*, Lecture Notes in Physics 972, https://doi.org/10.1007/978-3-030-51563-8_9

either a Higgs field, or a composite operator, in the adjoint representation of the gauge group. This approach has certain difficulties with N-ality, which I will come to. But let us begin with a success: Polyakov's demonstration, in 1975 [4], that magnetic monopoles are responsible for confinement in compact QED in $D = 3$ dimensions.

9.1   Magnetic Monopoles in Compact QED

Compact QED in three and four dimensions has monopole excitations, and these excitations are responsible for the confinement of electric charge. The confinement property exists only at strong lattice couplings in $D = 4$ dimensions, but it is found at all lattice couplings in $D = 3$ dimensions. The word "compact" refers to the compactness of the $U(1)$ gauge group in the lattice (as opposed to the continuum) formulation of electrodynamics. In $D = 3$ Euclidean dimensions it is sometimes useful to continue referring to $B_i(x) = \frac{1}{2}\varepsilon_{ijk}F_{jk}$ (rather than only the F_{12} component) as the "magnetic" field, and its integral over an area as the magnetic flux. I will follow that convention in this chapter.

Consider a plaquette variable $U(p) = \exp[i\theta(p)]$, i.e. the product of link variables around a plaquette, in $U(1)$ lattice gauge theory in $D = 3$ dimensions (compact QED_3). Obviously the lattice action is minimized for both magnetic flux zero, $\theta(p) = 0$, and magnetic flux $\theta(p) = 2\pi$. This double minimum is a reflection of the compactness of the gauge group in the lattice formulation. Suppose we have a series of parallel plaquettes with $\theta(p) = 2\pi$ arranged along, e.g., the positive x-axis. A line piercing this series of plaquettes can be thought of as an infinitely thin solenoid, known as a *"Dirac line"*, of flux 2π. At $x = 0$ this 2π flux cannot simply disappear, but must spread out through the plaquettes of a cube, one side of which has the incoming magnetic flux 2π (cf. Fig. 9.1). In other words, 2π flux enters through one plaquette of the cube, and then exits through the remaining sides of the cube. The flux which leaves the cube also adds up to 2π, but is less than that amount at any given plaquette. The 2π flux of the Dirac line is undetectable, and in fact the position of the Dirac line can be moved around via $U(1)$ gauge transformations, but the endpoint of the line is gauge invariant, and appears as a point-like source of magnetic field; i.e. a magnetic monopole. The Dirac line either

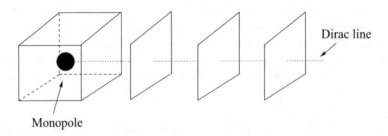

Fig. 9.1 In compact QED$_3$, a Dirac line is a line piercing the middle of plaquettes with flux $\Phi_B = 2\pi/e$. The line is bounded at one end by a Dirac monopole, and at the other by an antimonopole

goes off to infinity or, on a finite periodic lattice, ends in an antimonopole. In $D = 3$ dimensions monopoles are point-like objects, i.e. instantons. In $D = 4$ dimensions monopoles trace out trajectories, and are line-like objects.

Given a lattice configuration $U_\mu(x) = \exp[i\theta_\mu(x)]$, how do we locate the monopoles and determine their charges? The basic idea is to use Gauss's Law. Let's start in either $D = 3$ dimensions, or in a three-dimensional timeslice of a $D = 4$ dimensional lattice, and define the flux through a plaquette

$$f_{\mu\nu}(x) = \partial_\mu\theta_\nu(x) - \partial_\nu\theta_\mu(x) , \tag{9.1}$$

where, in this case, the partial derivatives must be defined as "forward" lattice derivatives, i.e. finite differences

$$\partial_\mu\theta_\nu(x) = \theta_\nu(x + \hat{\mu}) - \theta_\nu(x) . \tag{9.2}$$

Now if we would simply add up the flux around a cube (or any closed surface), it would just sum to zero as an identity. The trick is to remove from $f_{\mu\nu}(x)$ any flux which is due to a Dirac string. If that flux is removed, then the remaining magnetic flux is no longer necessarily divergence free. So, following De-Grand and Toussaint [5], let us express $f_{\mu\nu}$ as

$$f_{\mu\nu}(x) = \overline{f}_{\mu\nu}(x) + 2\pi n_{\mu\nu}(x) \quad , \quad \overline{f}_{\mu\nu}(x) \in [-\pi, \pi] , \tag{9.3}$$

where $n_{\mu\nu}(x)$ is a positive or negative integer. Let $\hat{e}(p; c)$ be the normal vector to plaquette p, pointing away from the center of cube c, and define $\Phi(p; c) = \frac{1}{2}\varepsilon_{ijk}e_i(p; c)\overline{f}_{jk}$. Then we obtain, from $\overline{f}_{\mu\nu}(x)$, the Gauss law for magnetic fields

$$2\pi m = \sum_{p \in c} \Phi(p; c)$$

$$= \tfrac{1}{2}\varepsilon_{ijk}\partial_i\overline{f}_{jk}(x) . \tag{9.4}$$

It is understood that the location of the magnetic charge m is at the center of the cube. The centers of all cubes on a $D = 3$ dimensional lattice are the sites of the dual lattice (cf. Sect. 6.2), and the magnetic monopoles are located at the sites of that structure.

It is natural to wonder if it is possible to express the action for compact QED$_3$ in terms of the magnetic monopoles and their interactions, integrating out all other degrees of freedoms. For this purpose, it is useful to use an approximation known as the *Villain* or periodic Gaussian model, in which the partition function is

$$Z = \int_{-\pi}^{\pi} \prod_{x,\mu} d\theta_\mu(x) \sum_{n_{\alpha\beta}(x)=-\infty}^{\infty} \exp\left[-\frac{1}{4g^2a} \sum_p (f_{\mu\nu}(x) - 2\pi n_{\mu\nu}(x))^2\right] ,$$

$$\tag{9.5}$$

where a is the lattice spacing, and $g_L^2 = g^2 a$ is the dimensionless lattice coupling in three dimensions. Like the Wilson action, the action in the Villain model is periodic with respect to $f_{\mu\nu} \rightarrow f_{\mu\nu} \pm 2\pi$, and in fact the Villain model approaches the usual Wilson action formulation as charge $g^2 \rightarrow 0$. In this model the integration over θ_μ can be carried out analytically, and after a number of steps that can be found in [4,6,7], one arrives at an action, for N monopoles, which involves only the monopole charges and their locations:

$$S_{mon} = \frac{2\pi^2}{g^2 a} \sum_{i \neq j} m_i m_j G(r_i - r_j) + \frac{2\pi^2}{g^2 a} G(0) \sum_i m_i^2 \,, \qquad (9.6)$$

where the indices i, j run from 1 to N, and $G(r - r')$ is the lattice Coulomb propagator in $D = 3$ dimensions

$$-\nabla_{lattice}^2 G(r - r') = \delta_{r,r'} \,, \qquad (9.7)$$

going as $1/4\pi|r - r'|$ at large separations. This expression was first derived by Polyakov [4, 6], and it represents the action of a gas of monopoles in three dimensions interacting via Coulombic forces; i.e. a monopole plasma. A Wilson loop, in this formulation, can be thought of as a current loop interacting with the magnetic field due to the monopole background. The loop expectation value can be calculated explicitly in $D = 3$ dimensions, and the result is that for a Wilson loop associated with n units of electric charge [4, 8]

$$\langle U_n(C) \rangle \approx \exp\left[-n\sigma \text{ area}(C)\right] \,, \qquad (9.8)$$

where string tension σ is a calculable function of coupling β, and the loop holonomy $U_n(C) = (U_1(C))^n$ in compact QED3.

A very rough image of what's going on is this: A Wilson loop can be thought of as a current loop which itself is the source of a magnetic field. In a monopole plasma, the free magnetic charges, i.e. the monopoles and antimonopoles, will tend to line up along the minimal area of the loop, and screen out the magnetic field that would have been generated by the Wilson (current) loop source (cf. Fig. 9.2). The area law falloff for Wilson loops is associated with the screening sheet of monopoles and antimonopoles along the minimal loop area.

In $D = 3$ dimensions, monopoles and antimonopoles form a plasma at any lattice coupling, rather than being bound in monopole-antimonopole pairs. This is because of the fact that, in three dimensions, the entropy of monopole location wins over the cost in monopole action. In $D = 4$ dimensions, monopoles trace out worldlines, and the creation and eventual annihilation of a monopole-antimonopole pair is associated with a closed worldline loop. At weak couplings, the cost in monopole action of a long loop exceeds the entropy associated with the size and shape of large loops. This means that there are only small loops, and in a timeslice one would find

Fig. 9.2 Monopoles and
antimonopoles in the plasma
line up along the minimal
area of the loop (solid line),
to screen the magnetic field
generated by the loop

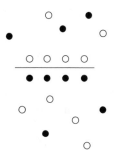

that every monopole was close to an antimonopole, with a separation on the order
of the typical extension of the small loops. In that case there is no monopole plasma
and no confinement. But as the coupling increases and the monople action $\sim 1/g^2$
decreases, there is a point where the entropy wins, and monopole loops percolate
through the volume. Now if we take a timeslice of the loop, the monopoles and
antimonopoles can be very far apart. The system is then in a plasma phase, and
confinement of electric charges is the result.

Polyakov's demonstration of confinement in compact QED$_3$ is quite beautiful
and it has been very influential, so the calculation is worth displaying (in the $n = 1$
single charge case) in a little more detail. First of all, at weak couplings monopoles
are very heavy, with action proportional to the square of magnetic charge, so
monopoles with charges $|m_i| > 1$ are suppressed. Then, allowing for the fact that
the number N of monopoles + antimonopoles is variable, the partition function for
a monopole Coulomb gas is

$$Z_{mon} = \sum_{N=0}^{\infty} \frac{\xi^N}{N!} \sum_{\{r_n\}} \sum_{\{m_n=\pm 1\}} \exp\left[-\frac{2\pi^2}{g^2 a} \sum_{i \neq j} m_i m_j G(r_i - r_j)\right], \qquad (9.9)$$

where

$$\xi = \exp\left[-\frac{2\pi^2}{g^2 a} G(0)\right] \qquad (9.10)$$

and $G(0) \approx 0.253$ in lattice units [9]. Introducing a field $\chi(r)$ to localize the action,

$$Z_{mon} = \int \prod_{r'} d\chi(r') \exp\left[-\frac{g^2 a}{4\pi^2} \sum_r \tfrac{1}{2}(\partial_\mu \chi(r))^2\right] \sum_{N=0}^{\infty} \frac{\xi^N}{N!} \sum_{\{r_n\}} \sum_{\{m_n=\pm 1\}} e^{i \sum_n m_n \chi(r_n)}$$

$$= \int \prod_{r'} d\chi(r') \exp\left[-\frac{g^2 a}{4\pi^2} \sum_r \tfrac{1}{2}(\partial_\mu \chi(r))^2\right] \sum_{N=0}^{\infty} \frac{\xi^N}{N!} \left[\sum_r 2\cos(\chi(r))\right]^N$$

$$= \int \prod_r d\chi(r) \exp\left[-\frac{g^2 a}{4\pi^2} \sum_r \left(\tfrac{1}{2}(\partial_\mu \chi)^2 - M_0^2 \cos \chi(r)\right)\right], \qquad (9.11)$$

where

$$M_0^2 = \frac{8\pi^2}{g^2 a} \exp\left[-\frac{2\pi^2}{g^2 a} G(0)\right] . \tag{9.12}$$

To make the expressions a little more compact, we switch to continuum notation. The translation is

$$(\partial_\mu)_{lattice} \to a\, (\partial_\mu)_{continuum}, \quad a^3 \sum_r \to \int d^3 r , \tag{9.13}$$

so that

$$Z_{mon} \approx \int D\chi(r)\, \exp\left[-\frac{g^2}{4\pi^2} \int d^3 r\, \left(\tfrac{1}{2}(\partial_\mu \chi)^2 - M^2 \cos \chi(r)\right)\right] , \tag{9.14}$$

where

$$M^2 = a^{-2} M_0^2 . \tag{9.15}$$

In the Coulomb gas approximation, the magnetic field due to the monopoles at sites $\{r_n\}$ can be obtained from the monopole density $\rho(r)$

$$\partial_\mu H_\mu(r) = 2\pi\, \rho(r), \quad \text{i.e.} \quad H_\mu(r) = \tfrac{1}{2} \int d^3 r'\, \frac{(r - r')_\mu}{|r - r'|^3}\, \rho(r') , \tag{9.16}$$

where

$$\rho(r) = \sum_{i=1}^{N} m_i \delta(r - r_i) \tag{9.17}$$

In particular we find, if C denotes a closed curve and $S(C)$ a surface with C as boundary, that

$$\oint_C dr_\mu A_\mu(r) = \int_{S(C)} dS_\mu(r)\, H_\mu(r) = \int d^3 r\, \eta_{S(C)}(r)\, \rho(r) , \tag{9.18}$$

where

$$\eta_{S(C)}(r) = -\tfrac{1}{2} \frac{\partial}{\partial r_\mu} \int_{S(C)} dS_\mu(r')\, \frac{1}{|r - r'|} . \tag{9.19}$$

If C is a planar curve in the (x, y) plane and $S(C)$ the planar surface with C as boundary curve, then we have

$$\partial^2 \eta_{S(C)} = 2\pi \delta'(z)\theta_{S(C)}(x, y) , \qquad (9.20)$$

where $\theta_{S(C)}(x, y) = 1$ for a point on $S(C)$ inside the boundary C, and $\theta_{S(C)}(x, y) = 0$ for a point outside the boundary C. For points close to the surface S

$$\eta_{S(C)}(r) = \pi \, \text{sign}(z) \, \theta_{S(C)}(x, y) , \qquad (9.21)$$

i.e. this quantity jumps by 2π across the surface.

A planar Wilson loop can now be expressed in terms of the monopole charge density, using (9.18), as follows:[2]

$$\langle U_1(C) \rangle \equiv \langle e^{i \oint dr_\mu \, A_\mu(r)} \rangle = \langle e^{i \int d^3r \, \eta_{S(C)}(r) \, \rho(r)} \rangle . \qquad (9.22)$$

Repeating the steps which lead from (9.9) to (9.14), inclusion of the Wilson loop (9.22) leads to a shift of $\chi(r)$ by $\eta_{S(C)}$, i.e.

$$\langle U_1(C) \rangle = \frac{1}{Z_{mon}} \int D\chi(r) \, \exp\left[-\frac{g^2}{4\pi} \int d^3r \, \left(\tfrac{1}{2}(\partial_\mu(\chi - \eta_{S(C)}))^2 - M^2 \cos \chi(r)\right)\right] . \qquad (9.23)$$

The dominant contribution to the expectation value (9.23) comes from the classical solution to the effective action in (9.23); the saddlepoint approximation is to estimate the functional integral by evaluating the exponent at the classical value. Choosing the planar loop to lie in the (x, y) plane we obtain from (9.20):

$$\partial^2 \chi = 2\pi \delta'(z)\theta_{S(C)}(x, y) + M^2 \sin \chi . \qquad (9.24)$$

The solution of this equation is discontinuous across the minimal surface of the loop, and at large M falls rapidly to zero away from the surface. Substituting this solution into the effective action results in an action density peaked in the neighborhood of the minimal area, and the increase in the action due to the Wilson loop source is proportional to the minimal area of the loop. In this way Polyakov obtained an the area law falloff for large Wilson loops

$$\langle U_1(C) \rangle \approx \exp\left[-\sigma \, \text{area}(C)\right] , \qquad (9.25)$$

from the three dimensional Coulomb gas of monopoles characteristic of compact QED in three Euclidean dimensions.[3]

---

[2] A factor of electric charge g has been absorbed into the definition of A_μ.

[3] A rigorous proof of confinement in compact QED$_3$ was later published in [10].

9.2   The Georgi-Glashow Model in $D = 3$ Dimensions

Confinement by monopoles in three-dimensional electrodynamics is a beautiful mechanism, but is it also relevant to non-abelian theories? The first place to look is in non-abelian theories, such as the Georgi-Glashow model, which have magnetic monopoles as solitonic solutions of the equations of motion.

The Georgi-Glashow model, also known as the $SU(2)$ adjoint Higgs model, is an $SU(2)$ gauge theory with a Higgs field in the adjoint representation. In $D = 3$ dimensions the action is

$$S = \int d^4x \left[\tfrac{1}{2}\text{Tr}[F_{\mu\nu}^2] + \tfrac{1}{2}(D_\mu \phi^a)^2 + \tfrac{1}{4}\lambda(\phi^a \phi^a - v^2)^2 \right] , \tag{9.26}$$

where ϕ^a has three color components, and D_μ is the usual covariant derivative. The potential is obviously minimized at $|\phi| = \sqrt{\phi^a \phi^a} = v$. Let us define

$$\widehat{\phi}^a = \frac{\phi^a}{|\phi|} . \tag{9.27}$$

If we transform to a unitary gauge, where $\widehat{\phi} = \delta^{a3}$, then there is still a residual $U(1)$ local gauge invariance, which leaves $\widehat{\phi}$ invariant. It is tempting, in this gauge, to regard the model as a gauge theory of "photons," i.e. the gauge potentials A_μ^3, which are neutral with respect to the $U(1)$ subgroup, coupled to electrically charged "W-bosons" $W_\mu^\pm = (A_\mu^1 \mp i A_\mu^2)/\sqrt{2}$. The corresponding abelian field strength due to the photons would be $f_{\mu\nu}^3 = \partial_\mu A_\nu^3 - \partial_\nu A_\mu^3$. 't Hooft [11] found a gauge invariant field strength

$$\begin{aligned}
\mathscr{F}_{\mu\nu} &= \partial_\mu(\widehat{\phi}^a A_\nu^a) - \partial_\nu(\widehat{\phi}^a A_\mu^a) - \frac{1}{g}\varepsilon^{abc}\widehat{\phi}^a \partial_\mu\widehat{\phi}^b \partial_\nu\widehat{\phi}^c \\
&= F_{\mu\nu}^a \widehat{\phi}^a - \frac{1}{g}\varepsilon^{abc}\widehat{\phi}^a D_\mu\widehat{\phi}^b D_\nu\widehat{\phi}^c ,
\end{aligned} \tag{9.28}$$

which obviously reduces to the abelian field strength in unitary gauge. This tensor is used to define the electromagnetic field strength gauge-invariantly in the Georgi-Glashow model.

't Hooft [11] and Polyakov [12] found that the equations of motion of the Georgi-Glashow model contain magnetic monopole solutions which are instantons, in $D = 3$ dimensions, or static solutions, in $D = 3 + 1$ dimensions. A monopole centered at the origin has, in a convenient gauge, the "hedgehog" form

$$\phi^a(x) = v\frac{x^a}{r}f(r) ,$$

$$A_i^a(x) = \varepsilon_{aij}\frac{x^j}{gr^2}a(r) , \tag{9.29}$$

where $f(r), a(r)$ satisfy certain differential equations, with $f(0) = a(0) = 0$, and $f(\infty) = a(\infty) = 1$. This solution can be identified as a magnetic monopole because the abelian magnetic field, derived from the electromagnetic field strength $\mathcal{F}_{ij}$, has the monopole form at large r

$$B_i = \tfrac{1}{2}\varepsilon_{ijk}\mathcal{F}_{jk} \rightarrow -\frac{1}{g}\frac{x^i}{r^3}\,, \qquad (9.30)$$

so that the surface integral of the B-field, evaluated on a large closed surface (e.g. a sphere at large r) enclosing the origin will be

$$\int d\mathbf{S} \cdot \mathbf{B} = -\frac{4\pi}{g}\,. \qquad (9.31)$$

Since the flux integrated over a closed surface is equal (Heaviside-Lorentz convention) to the charge contained inside, we conclude that the magnetic charge of the magnetic monopole is $-4\pi/g$. The mass of a magnetic monopole, in $D = 3 + 1$ dimensions, is

$$M_{mon} = \frac{4\pi v}{g}h\left(\frac{\lambda}{g^2}\right)\,, \qquad (9.32)$$

where $h(x)$ is a slowly varying function of order one.

In a unitary gauge, where ϕ^a is rotated to point in the positive color 3 direction, the A_μ^3 field is the photon, and the charged W bosons are the $W_\mu^\pm = (A_\mu^1 \mp iA_\mu^2)/\sqrt{2}$ fields. In this gauge the 't Hooft-Polyakov monopole lies in the color 3 direction, and has the form of a Dirac monopole at large r. In other words, if we consider a surface enclosing the monopole, then there is a physically undetectable Dirac string, carrying a magnetic flux of $4\pi/g$ in the color 3 direction, which enters the surface, and an equal magnetic flux, diverging from the center of the monopole, which leaves the surface. As in compact QED$_3$, the location of the Dirac string can be varied by a gauge transformation, although its endpoint on the monopole is fixed. A point to note is that the Higgs field vanishes at the center of the monopole in any gauge; this will be important when we come to the abelian projection proposal for identifying monopoles in pure gauge theories.

't Hooft-Polyakov monopoles again interact via a Coulombic potential, and the partition function for the corresponding monopole Coulomb gas can be brought to the form of Eq. (9.14), except that now the parameter M, which diverges in compact QED$_3$ in the lattice spacing $a \rightarrow 0$ limit, is replaced by a finite constant, dependent on the coupling g and W-boson mass. Wilson loops can be computed exactly as in compact QED$_3$, and a finite string tension $\sigma \sim \exp[-S_{mon}]$ is obtained [6]. It would *seem* that in the Georgi-Glashow model, the goal of separating the gauge field into a piece $\mathcal{A}$, which carries only the confining fluctuations, and the remaining non-confining fluctuations $\widetilde{A} = A - \mathcal{A}$ has been accomplished unambiguously: The instruction is to go to unitary gauge, identify the monopole positions $\{x_k^\mu\}$ from the zeros of the Higgs fields, and let $\mathscr{A}^a_\mu(x) = \delta^{a3}\mathscr{A}_\mu$ be the vector potential of Dirac monopoles at those positions. The corresponding long-range magnetic field at point x, due to monopoles and antimonopoles centered at points $\{x_k\}$ is then

$$B^3_\mu(x) = \sum_k q_k \frac{(x - x_k)_\mu}{|\mathbf{x} - \mathbf{x}_k|^3} , \tag{9.33}$$

where $q_k = \pm 1/g$. This identification of the confining fluctuations is certainly correct for compact QED$_3$. It is a little problematic, however, in the Georgi-Glashow model.

The problem is connected with loops of multiple electric charge. Consider, in unitary gauge, loops of the form

$$U_n(C) = \exp\left[in\frac{1}{2} \oint_C A^3_\mu(x)dx^\mu \right] . \tag{9.34}$$

If the confining fields are those generated by a monopole Coulomb gas, then the charge-dependence of the string tension is the same as for compact QED$_3$, namely

$$\langle U_n(C) \rangle \sim e^{-n\sigma A(C)} . \tag{9.35}$$

with $A(C)$ the minimal area. But this answer cannot be correct in the Georgi-Glashow model, as pointed out in Ref. [8], because there are charged $W^\pm$ bosons carrying two units of the minimum electric charge. This means that static charges carrying n units of electric charge are screened to n mod 2 units of charge, and therefore

$$\langle U_n(C) \rangle \sim \begin{cases} e^{-\sigma A(C)} & n \text{ odd} \\ e^{-\mu P(C)} & n \text{ even} \end{cases} . \tag{9.36}$$

The vector potential obtained from the monopole Coulomb gas action leads to the wrong answer for these observables, and from this it follows that confining magnetic flux derived from $\mathscr{A}^a_\mu(x)$ is not distributed in accordance with a monopole plasma, at least, not at very large distance scales (we will expand on this point in Sect. 9.6.1).

What we learn from this is that the infrared effects of W bosons cannot be neglected in constructing an effective long-range action for the Georgi-Glashow theory. The usual justification for ignoring the W bosons is that these are massive objects, and therefore cannot affect the dynamics at large scales. For a confining theory, that argument is simply wrong. Moreover, confining fluctuations must organize themselves at large scales in such a way that $n = $ even charged loops have a vanishing string tension. The most obvious way to satisfy this condition is for the confining flux to be collimated into Z_2 vortices, rather than being distributed as in a monopole Coulomb gas [8].

On the other hand, the monopole Coulomb gas approximation ought to have *some* range of validity in the $D = 3$ Georgi-Glashow model. On the basis of simple energetics arguments, the electric flux tube between static sources carrying two units of $U(1)$ electric charge will break by W pair creation only when the energy of the flux tube is greater than the combined mass of the two W bosons; i.e. when $L\sigma_2 > 2m_W$. The double-charge string tension σ_2 is zero beyond the string-breaking distance; however, up to that point, the monopole Coulomb gas result that $\sigma_2 = 2\sigma_1$ should hold. Then the string-breaking distance is $L \approx m_W/\sigma_1$, and up to that length scale a monopole Coulomb gas analysis ought to give a good account of the dynamics.

What about $D = 4$ dimensions? In four dimensions the Georgi-Glashow model has a confining phase and a non-confining phase [13–15], and only in the non-confining phase, where it is possible to perturb (in, e.g., unitary gauge) around a non-zero value of the scalar field, are there stable magnetic monopole solutions. Monopoles and antimonopoles trace out worldlines in four dimensions, and monopole antimonopole pairs trace out closed loops. In the non-confining phase these loops are small and do not percolate (in the Euclidean theory) throughout the spacetime volume. In a time slice, one would find that every monopole was paired with an antimonopole. The monopoles do not form a plasma, and it appears that the Polyakov calculation is not applicable.

A possible way to extend the $D = 3$ physics a little way into $D = 4$ dimensions is to consider a pure gauge theory in a four-dimensional space in which one of the space dimensions is compactified, i.e. we apply periodic boundary conditions in one of the dimensions, so space is not R^4 but $R^3 \times S^1$. Polyakov lines through the compact direction then play a role somewhat analogous to the Higgs field in the Georgi-Glashow model. However, the expectation value of the trace of the Polyakov line is non-zero if the radius of S^1 is small, so from the 4D viewpoint the compactified model is in the deconfined phase, and a phase transition separates the large and small radius regimes. It has been pointed out by Shifman and Ünsal [16] that this deconfinement transition could be avoided by adding to the action of an $SU(N)$ gauge theory the following term

$$\Delta S = \int_{R^3} d^3x \, \frac{2}{\pi L^3} \sum_{n=1}^{[N/2]} d_n \left| \text{Tr} U^n(\mathbf{x}) \right|^2 , \qquad (9.37)$$

where L is the extension of the compact direction, $U(\mathbf{x})$ is a Polyakov loop holonomy winding around the compact direction, $\mathbf{x}$ denotes coordinates in R^3, and $[N/2]$ is the integer part of quantity in brackets. For small L and judiciously chosen values d_n, the eigenvalues of the Polyakov line $U(\mathbf{x})$ are driven as far apart as possible, so that the action is minimized at $P(\mathbf{x}) = \text{Tr} U(\mathbf{x}) = 0$, and center symmetry is unbroken. It is then possible to demonstrate confinement via a monopole mechanism, along the same lines as the $D = 3$ Georgi-Glashow model. At large L, of course, center symmetry arises in a rather different fashion; it is due to large and random fluctuations in the eigenvalues of the Polyakov line. At large L there is no obvious abelian confinement mechanism, and the analytical control available at small L is lost. Nevertheless, the Shifman-Ünsal approach does demonstrate confinement in non-supersymmetric theories in $D = 4$ dimensions for small L, and L can be regarded as a parameter which may interpolate continuously between an abelian monopole mechanism, and (since ΔS is negligible at large L) whatever mechanism is operative in R^4. The approach has been extended to theories with fermions in various representations. In fact, with massless fermions in the adjoint representation, ΔS can be dispensed with, because a similar effect is found at the level of the effective action, obtained after integrating out the massive Kaluza-Klein excitations that arise due to the compact direction [17]. It is interesting that in some theories of that kind, which includes $\mathcal{N} = 1$ supersymmetric Yang-Mills theory, there are two kinds of magnetic monopoles: the usual 't Hooft-Polyakov monopoles, and "Kaluza-Klein" monopoles associated with the fact that one of the spatial directions is periodic and small. It can be shown that these two types of monopoles can form bound states with two units of magnetic charge, but zero topological charge, which are known as "bions." Confinement in these theories is attributed to the condensation of bions, rather than isolated monopoles with non-zero topological charge [17].

9.3   Dual Superconductivity, and the Seiberg-Witten Model

When a type II superconductor is placed in an external magnetic field, the magnetic field can only penetrate the superconductor in cylindrical regions known as Abrikosov vortices, which carry only a certain fixed amount of magnetic flux. Abrikosov vortices are magnetic flux tubes, and in principle such a flux tube could begin at a monopole, of appropriate magnetic charge, and end on an antimonopole. The constant energy density along an Abrikosov vortex then implies a linear potential between the monopole and antimonopole. In other words, a type II superconductor, which is a system in which bosons (the Cooper pairs) of electric charge $2e$ are condensed, is a system which squeezes magnetic fields into tubes of quantized flux, and confines magnetic charge via a linear potential. So if there *were* real magnetic monopoles and antimonopoles present in a material of this kind, we would observe that these objects were confined.

The example of type II superconductors becomes even more suggestive in view of the electric/magnetic *duality* symmetry of Maxwell's equations. If we are willing to allow for the existence of elementary magnetic charges g_M, then the corresponding Maxwell's equations

$$\partial^\mu F_{\mu\nu} = j_\nu^e \ , \quad \partial^{\mu*}F_{\mu\nu} = j_\nu^m \tag{9.38}$$

are symmetric with respect to the interchange of fields $F_{\mu\nu} \rightarrow {}^*F_{\mu\nu}, {}^*F_{\mu\nu} \rightarrow -F_{\mu\nu}$, and currents, i.e.

$$\mathbf{E} \rightarrow \mathbf{B} \ , \quad \mathbf{B} \rightarrow -\mathbf{E} \ , \quad j_\mu^e \leftrightarrow j_\mu^m \ , \quad j_\mu^m \leftrightarrow -j_\mu^e \ , \tag{9.39}$$

where $^*F_{\mu\nu}$ is the dual field strength tensor

$$^*F_{\mu\nu} = \tfrac{1}{2}\varepsilon^{\mu\nu\alpha\beta}F_{\alpha\beta} \ . \tag{9.40}$$

Conservation of both the electric and magnetic currents implies separate electric and magnetic $U(1)$ symmetries. It occurred independently to 't Hooft [2] and Mandelstam [3], in 1976, that quark confinement in QCD might be due to some type of *dual superconductivity*, in which the QCD vacuum state can be regarded as a condensate of a magnetically charged boson field (magnetic monopoles), confining electrically charged particles (the quarks). In other words, a dual superconductor is like an ordinary superconductor in which the roles of the **E** and **B** fields, and electric and magnetic charges, have been interchanged. This approach entails, as in the Georgi-Glashow model, singling out an abelian subgroup of the full gauge group.

The idea is expressed more concretely in a relativistic generalization of the Landau-Ginzburg theory of superconductivity. The relativistic action is the *abelian Higgs model*

$$S = \int d^D x \left(\frac{1}{4}F_{\mu\nu}F^{\mu\nu} + \left|(\partial_\mu + ieA_\mu)\varphi\right|^2 + \frac{\lambda}{4}(\varphi^*\varphi - v^2)^2 \right) . \tag{9.41}$$

The abelian Higgs model has magnetic flux-tube solutions analogous to Abrikosov vortices, which are known as *Nielsen-Olesen vortices* [18]. For a vortex along the z-axis, the Nielsen-Olesen vortex in cylindrical (r, ϕ, z) coordinates has the following asymptotic form:

$$\varphi(r, \phi) = v f(r)e^{in\phi}$$
$$A_\phi(r) = \frac{n}{er}a(r) , \tag{9.42}$$

with $f(0) = a(0) = 0$ and $f(\infty) = a(\infty) = 1$. The topological stability of these vortices is related to the fact that the Higgs potential has its minimum away from $\varphi = 0$, and that the complex phase of the Higgs field along a closed loop around the vortex has a non-zero integer winding number n. The magnetic flux carried by a Nielsen-Olesen vortex is $2\pi n/e$. In the dual version of the abelian Higgs model, the Higgs field is magnetically charged, and it couples, not to the usual vector potential A_μ, but rather to a "dual" photon field C_μ whose associated field strength is $^*F_{\mu\nu}$. Then the Nielsen-Olesen vortex is an electric, rather than magnetic flux tube, and it is electric charge which is confined.

The dual-superconductor idea is not only very attractive, but it can also be demonstrated to be operating in a particular theory: the Seiberg-Witten model. The important work of Seiberg and Witten on duality and confinement in supersymmetric gauge theories is beyond the scope of this book, and the reader is referred to the original articles [19, 20] and reviews [21–23] for an exposition. A few comments, however, are relevant at this point.

The Seiberg-Witten work is concerned with supersymmetric Yang-Mills theory. Supersymmetry is an extension of Poincare symmetry, adding to the usual group generators (translations, rotations, and boosts) of the Poincare Lie algebra a set of $\mathcal{N}$ supersymmetry generators. These enlarge the Lie algebra to include certain anticommutator relations among the supersymmetry generators. The number $\mathcal{N}$ of different supersymmetry generators is, in general, a power (≥ 0) of 2. Irreducible representations of supersymmetry contain both bosonic and fermionic particle states, with equal numbers of bosonic and fermionic degrees of freedom in the supermultiplet. The (bosonic + fermionic) field content of a supersymmetric field theory, as well as the interaction terms in such theories, are highly constrained.

$\mathcal{N} = 2$ super Yang-Mills theory, like the Georgi-Glashow model, has scalar fields in the adjoint representation of the gauge group, which can be used to single out a compact abelian subgroup, which is $U(1)^{N-1}$ for the $SU(N)$ gauge group. The theory does not have a unique vacuum; rather there is a continuous set of inequivalent vacua, which are specified by the values ("moduli") of certain scalar field operators. Seiberg and Witten were able to show that there is a point in moduli space at which the magnetic monopoles of this theory become massless. On adding a soft supersymmetry breaking term, which reduces the $\mathcal{N} = 2$ supersymmetry to $\mathcal{N} = 1$, the theory goes into a confining phase associated with a condensation of the monopole field. Seiberg and Witten derive in this case an effective low energy action which is a supersymmetric generalization of the dual abelian Higgs model. This is a truly remarkable result: the dual abelian Higgs model is expressed in terms of dual (magnetically charged) fields, even though such fields were invisible in the Lagrangian of the starting point, which is $\mathcal{N} = 2$ super Yang-Mills theory. Once a dual abelian Higgs model has been obtained, electric flux tube formation and confinement of electric charge follows, along the same lines as magnetic flux tube confinement of magnetic charge in the ordinary abelian Higgs model, and in type II superconductors. The Sine Law rule for k-string tensions, already mentioned in Chap. 5, was motivated by the $SU(N)$ version of the Seiberg-Witten model, where the Sine Law was derived by Douglas and Shenker [24]. These authors also pointed out that the $SU(N)$ gauge group leads to $N - 1$ separate copies of the dual abelian Higgs model, and thus $N - 1$ distinct types of electric flux tube, which is not what one expects in a pure $SU(N)$ gauge theory.

The validity of the effective low energy theory at very large distance scales is questionable, however. A flux tube will also form between objects which carry two units of the $U(1)$ electric charge, and there is nothing in the effective dual abelian Higgs model which would allow that flux tube to break. On the other hand, in the $SU(2)$ theory these electric charges have zero N-ality, and the flux tube *must* break at some point, in this case via pair creation of W-bosons. Just as in Polyakov's monopole gas treatment of the $D = 3$ dimensional Georgi-Glashow model, the Seiberg-Witten low-energy effective action neglects the W bosons of the theory, on the grounds that they are massive and should not contribute to long-range physics. But we have already seen that the W bosons are very relevant to long-distance physics; the N-ality dependence of string tensions σ_r cannot be obtained without them. This means that the Seiberg-Witten effective action cannot be the whole story at sufficiently large distance scales. To put it another way: the "low-energy" effective action is not necessarily the long-distance effective action. In the Seiberg-Witten theory, the effective action is obtained keeping only local terms with no more than two derivatives of the bosonic fields. It is not obtained by actually integrating out the massive W bosons, for otherwise the effective action would have the correct N-ality dependence built in. At distance scales on the order of the color screening length, $L \approx m_W/\sigma$, the dual abelian Higgs action arrives at the wrong N-ality dependence for abelian Wilson loops, and on those grounds it is inadequate to describe the large-scale vacuum fluctuations in the theory. There is, as in the Georgi-Glashow model, an intermediate range of distances up to the screening scale for which the Seiberg-Witten effective action gives a good account of the physics. Beyond that scale the contributions of virtual W bosons are important, and can no longer be ignored.

9.4 The Abelian Projection

The example of monopole confinement in compact QED, in the Georgi-Glashow model, and in the Seiberg-Witten model, raises the hope that some similar mechanism might exist in pure gauge theories, or gauge theories without fundamental matter fields. To make this idea work, one has to identify an abelian subgroup of the gauge group, associated with the abelian magnetic monopole field. Since there is no elementary Higgs field in the adjoint representation, whose unitary gauge would single out an appropriate abelian subgroup, 't Hooft [25] suggested that a composite operator involving only gluonic fields, and behaving under a gauge transformation $G(x)$ like a matter field in the adjoint representation, might serve the same purpose. For a matrix-valued operator $X(x)$, the gauge transformation law would be

$$X(x) \rightarrow G(x)X(x)G^{\dagger}(x) . \tag{9.43}$$

An example of such an operator would be any spatial component of the non-abelian field strength tensor, e.g. $F_{12}(x)$. The analog of a unitary gauge would be the gauge in which the operator X is diagonal. In an $SU(N)$ theory X is an $N \times N$ matrix, and in the gauge where

$$X = \text{diag}[\lambda_1, \lambda_2, \ldots, \lambda_N] \quad \text{with} \quad \lambda_1 < \lambda_2 < \ldots < \lambda_N \tag{9.44}$$

there is left unfixed an abelian subgroup $U(1)^{N-1}$, whose generators belong to the Cartan subalgebra of the gauge group. The Cartan subalgebra is the largest subset of group generators which commute with one another, but the choice is not unique. For example, the generators of $SU(2)$ are the three Pauli matrices ($\times \frac{1}{2}$). The $U(1)$ subgroup generated by any one of the Pauli matrices can be taken as the Cartan subgroup. For $SU(3)$, one could choose, e.g., the third component of isospin I_3, and hypercharge Y, forming the subgroup $U(1) \times U(1)$. In general, for any $SU(N)$ group, the Cartan subgroup is $U(1)^{N-1}$.

In a gauge in which the operator X is diagonal, the gauge theory can be thought of as an abelian $U(1)^{N-1}$ gauge theory, with "photon" and monopole excitations, the photons corresponding to the generators of the Cartan subalgebra. In addition there are "matter" fields (the remaining gluons), which are electrically charged with respect to the Cartan subgroup. The idea is that confinement is due to monopole condensation, as in the dual abelian Higgs model. The locations of the monopoles are taken to be at those sites where the unitary gauge fixing is ambiguous, and this is where two of the eigenvalues of the X operator coincide (recall that the Higgs field, which is used to define the unitary gauge, vanishes at the center of a 't Hooft-Polyakov monopole). Of course, the locations of such monopoles are entirely dependent on the choice of operator X; it is not at all obvious that there is any real physical content associated with points or worldlines located in this way. Indeed, the gauge based on the operator $X = F_{12}$ has not led to anything interesting.

Gauges which fix the gauge symmetry only up to the residual symmetry of the Cartan subgroup are known as *abelian projection gauges*, and they do not necessarily require that we know the operator X explicitly. In particular, since the goal is to identify an abelian confinement mechanism inside a non-abelian gauge theory, it is reasonable that the idea might work best, in an $SU(N)$ gauge theory, in a gauge where the theory looks most like a $U(1)^{N-1}$ gauge theory. This would be a gauge where the quantum fluctuations of "charged" gluon degrees of freedom are much suppressed, compared to the quantum fluctuations of the "photon" degrees of freedom. So we look for a gauge in which the link variables are as diagonal as possible. This is known as the *maximal abelian gauge* [26], and in $SU(2)$ gauge theory, it is specified by the condition that

$$R = \sum_{x,\mu} \text{Tr}[U_\mu(x)\sigma_3 U_\mu^\dagger(x)\sigma_3] \tag{9.45}$$

is maximized.

Fixing to the maximal abelian gauge leaves a residual $U(1)$ gauge symmetry

$$U_\mu(x) \to e^{i\phi(x)\sigma_3} U_\mu(x) e^{-i\phi(x+\hat{\mu})\sigma_3} . \tag{9.46}$$

Let $u_\mu(x)$ be the diagonal part of $U_\mu(x)$, rescaled to restore unitarity

$$U_\mu(x) = a_0 \mathbb{1} + i\mathbf{a} \cdot \boldsymbol{\sigma}$$

$$u_\mu(x) = \frac{1}{\sqrt{a_0^2 + a_3^2}}\left[a_0 \mathbb{1} + i a_3 \sigma_3\right]$$

$$= \begin{bmatrix} e^{i\theta_\mu(x)} & \\ & e^{-i\theta_\mu(x)} \end{bmatrix} . \tag{9.47}$$

Then, make the decomposition

$$U_\mu(x) = C_\mu u_\mu(x)$$

$$= \left[\begin{array}{cc} \left(1 - |c_\mu(x)|^2\right)^{1/2} & c_\mu(x) \\ -c_\mu^*(x) & \left(1 - |c_\mu(x)|^2\right)^{1/2} \end{array} \right] \left[\begin{array}{cc} e^{i\theta_\mu(x)} & \\ & e^{-i\theta_\mu(x)} \end{array} \right] \quad (9.48)$$

What is interesting is that under the remnant $U(1)$ gauge symmetry, $\theta_\mu(x)$ transforms like an abelian gauge field, and $c_\mu(x)$ transforms like a matter field carrying two units of the abelian charge, i.e.

$$e^{i\theta_\mu(x)} \rightarrow e^{i\phi(x)} e^{i\theta_\mu(x)} e^{-i\phi(x+\hat{\mu})}$$

$$c_\mu(x) \rightarrow e^{2i\phi(x)} c_\mu(x) , \quad (9.49)$$

so the particle content of the theory in maximal abelian gauge consists of "photons" coupled to double-charged matter fields. The *abelian projected lattice* is obtained by the replacement $U_\mu(x) \rightarrow \exp[i\theta_\mu(x)]$.

The $U(1)$ gauge symmetry is compact, and there are monopole currents $k_\mu(x)$ in the theory, which can be extracted from the abelian-projected links $\exp[i\theta_\mu(x)]$, using the generalization, to $D = 4$ dimensions, of $(9.4)^4$

$$k_\mu(x) = \frac{1}{4\pi} \varepsilon_{\mu\alpha\beta\gamma} \partial_\alpha \overline{f}_{\beta\gamma}(x) . \quad (9.50)$$

One can then construct link variables from the Coulombic fields of those monopoles alone, discarding the "photon" contributions, just from the Dirac string variables $n_{\mu\nu}$ [27]

$$u_\mu^{mon}(x) = \exp[i\theta_\mu^{mon}(x)]$$

$$\theta_\mu^{mon}(x) = -2\pi \sum_y G(x - y)\partial_\nu' n_{\mu\nu}(y) , \quad (9.51)$$

where $G(x - y)$ is again the lattice Coulomb propagator (inverse of the lattice Laplacian), and this time ∂_μ' denotes the backward lattice derivative. Replacing link variables by the $u_\mu^{mon}(x)$, and calculating observables (such as Wilson loops) in that ensemble, is known as the *monopole dominance* approximation [28, 29].

One suggested application of abelian projection is to help define a monopole creation operator $\mu(\mathbf{x})$, whose expectation value could serve as an order parameter for dual superconductivity in a non-abelian gauge theory [30, 31]. Recall that in a

---

[4]The monopole currents actually live on the links of the dual lattice (cf. Sect. 6.2) in four dimensions.

compact $U(1)$ gauge theory the conserved magnetic current is associated with a dual $U(1)$ gauge symmetry, and the total magnetic charge generates a global subgroup of this dual gauge symmetry. If we define an operator $\mu(\mathbf{x})$ which acts on states in the Schrodinger representation by inserting a monopole field configuration $A_i^M(\mathbf{y})$ centered at the point $\mathbf{y} = \mathbf{x}$, i.e.

$$\mu(\mathbf{x})|A_i\rangle = |A_i + A_i^M\rangle , \qquad (9.52)$$

then μ does not commute with the total magnetic charge, and $\langle \mu \rangle \neq 0$ is a signal that the associated global symmetry is spontaneously broken. In a non-abelian gauge theory, an abelian projection gauge is introduced to single out an abelian $U(1)^{N-1}$ subgroup, and μ is defined in terms of the gauge fields associated with that abelian subgroup. If the expectation value of μ is non-zero, it means that the system is in the phase of "dual superconductivity." The proposal is that a gauge theory is in a confinement phase if and only if $\langle \mu \rangle > 0$, and a phase transition to $\langle \mu \rangle = 0$ is therefore a transition from the confined to the deconfined phase. In fact, in very many cases it has been shown numerically that the high-temperature deconfinement transition does coincide with $\langle \mu \rangle \to 0$. The problem with the proposal, however, is that the transition $\langle \mu \rangle \to 0$ is also found in cases where there is a bulk transition that is *not* associated with a transition to the deconfined phase, and there are also cases where the transition $\langle \mu \rangle \to 0$ occurs in the absence of any thermodynamic transition whatever [32].

9.4.1 Monopoles and Vortices

The monopole dominance approximation has been investigated numerically [27–29] in SU(2) lattice gauge theory, and in this approximation the string tensions come out about right (i.e. almost equal to the tension of gauge invariant loops) for single charged ($n = 1$) Wilson loops. We write

$$W_n^{abel}(C) = \left\langle \exp[in\theta(C)] \right\rangle ,$$
$$W_n^{mon}(C) = \left\langle \exp[in\theta^{mon}(C)] \right\rangle . \qquad (9.53)$$

as the Wilson loops corresponding to n units of electric charge in the abelian projected lattice $U_\mu(x) \to \exp[i\theta_\mu(x)]$, and in the monopole dominance approximation $U_\mu(x) \to \exp[i\theta_\mu^{mon}(x)]$, respectively, where $\theta(C)$ is the lattice "loop integral" of link angles $\theta_\mu(x)$ around the loop C. Because of screening by the double-charged (off-diagonal) gluons, we should have

$$\sigma_n = \begin{cases} \sigma_1 & n \text{ odd} \\ 0 & n \text{ even} \end{cases} . \qquad (9.54)$$

However, charge screening is difficult to check numerically in Wilson loop operators. Instead, the expectation values of double-charged Polyakov loops have been computed in both the abelian projection, and the monopole dominance approximation. Because of charge-screening, we expect

$$\langle P_n(\mathbf{x}) \rangle = \begin{cases} = 0 & n \text{ odd} \\ \neq 0 & n \text{ even} \end{cases}. \tag{9.55}$$

For single-charged $n = 1$ loops, below the deconfinement transition temperature, both the abelian projected and monopole dominance Polyakov loops have expectation values consistent with zero, as they should. However, the abelian projection and monopole dominance approximation give drastically different results for double-charged loops [33]. For the $n = 2$ loops, the expectation values of abelian-projection Polyakov loops are non-zero, as they should be. In contrast, the corresponding expectation values in the monopole dominance approximation are found to be consistent with zero, and this conflicts with the requirements of N-ality, and the existence of charge screening. It follows that, even if the confining disorder is dominated (in some gauge) by abelian configurations, the distribution of abelian flux *cannot* be that of a monopole Coulomb gas; such a distribution disorders both $n = 1$ *and* $n = 2$ loops. A similar argument applies to dual superconductors: if there is confinement for $n = 1$ electrically charged objects, then $n = 2$ charge is also confined, and this is incompatible with the existence of charge screening. In a type-II dual superconductor there would simply be two, rather than one, electric flux tubes running between doubly-charged sources.

On the other hand, the abelian projection does get some things right, such as the string tension of the fundamental representation Wilson loop on an abelian projected lattice [27, 34]. To see what may be going on, let's think of how vortices would look in maximal abelian gauge, at some fixed time t. Figure 9.3 is a sketch of a thick center vortex in the absence of gauge-fixing. In this case the vortex field strength $F_{\mu\nu}^a$ points in random directions in the Lie algebra. Fixing to maximal abelian gauge, the field tends to line up in the $\pm\sigma_3$ direction, but there will still be regions where the field strength rotates in the group manifold, from $+\sigma_3$ to $-\sigma_3$ (Fig. 9.4). Now, if we keep only the abelian part of the link variables $u_\mu(x)$ (i.e. abelian projection), we get a monopole-antimonopole chain (Fig. 9.5), with $\pm\pi$ magnetic flux running between a monopole and neighboring antimonopole. The total monopole flux is $\pm 2\pi$, as it should be. Then a typical vacuum configuration at a fixed time, after abelian projection, looks something like the sketch in Fig. 9.6. Double-charged Wilson loops and Polyakov loops, in the abelian projection, are

Fig. 9.3 Vortex field strength before gauge fixing. The arrows indicate direction in color space

Fig. 9.4 Vortex field strength after maximal abelian gauge fixing. Vortex strength is mainly in the $\pm\sigma_3$ direction

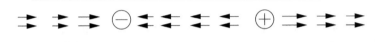

Fig. 9.5 Vortex field after abelian projection

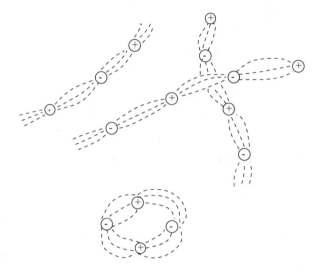

Fig. 9.6 Hypothetical collimation of monopole/antimonopole flux into center vortex tubes on the abelian-projected lattice

insensitive to linking with such vortices, and this will ensure that the string tensions σ_n will satisfy (9.54) as required.

The picture outlined above can be tested by numerical simulation. We work in the indirect maximal center gauge, which uses maximal abelian gauge as an intermediate step, and identify the locations of both monopoles (by abelian projection) and vortices (by center projection). We also measure the excess action (above the average plaquette value S_0), on plaquettes belonging to monopole 'cubes', and on plaquettes pierced by vortex lines. The following results are obtained for $SU(2)$ lattice gauge theory at $\beta = 2.4$ [33]:

1. Almost all monopoles and antimonopoles (97%) lie on vortex sheets.
2. At fixed time, the monopoles and antimonopoles alternate on the vortex lines, in a chain.

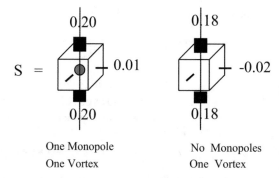

One Monopole · One Vortex

No Monopoles · One Vortex

Fig. 9.7 The left figure shows the plaquette excess action distribution on an isolated monopole cube (no nearest neighbor monopole currents) pierced by a single P-vortex. In the illustration the bottom and top plaquettes are pierced by the vortex. For comparison, the excess action distribution is also shown for a no-monopole cube (right figure) pierced by a P-vortex. The simulation is at $\beta = 2.4$ in $SU(2)$ lattice gauge theory. From Ambjørn et al. [33]

3. Excess (gauge-invariant) plaquette action is highly directional (cf. also Gubarev et al. [35]), and lies mainly on plaquettes pierced by vortex lines as indicated in Fig. 9.7. The presence or absence of a monopole in the cube is not so important to the excess action. Very similar results are obtained from 2^3 and 3^3 cubes.

9.5 Calorons

As we saw earlier in this chapter, Polyakov managed to derive confinement analytically in compact QED_3, by summing over classical solutions, in this case monopoles, of the Euclidean field equations. Monopoles are instantons in compact QED_3, and it is natural to wonder if confinement could be derived from some semiclassical treatment of Yang-Mills theory based on the instanton solutions of non-abelian gauge theories. The standard instantons, introduced by Belavin et al. [36], do not seem to work; their field strengths fall off too rapidly to produce the desired magnetic disorder in the vacuum. In recent years, however, it has been realized that instanton solutions at finite temperature, known as *calorons*, might do the job. These caloron solutions were introduced independently by Kraan and van Baal [37,38] and Lee and Lu [39] (KvBLL), and they have the remarkable property of containing monopole constituents which may, depending on the type of caloron, be widely separated.

The monopole constituents of the KvBLL calorons are also called *dyons* because they are a source of both electric and magnetic fields. For an $SU(N)$ gauge group there are N types of dyons, and in a suitable gauge, far from the dyon center, their field strengths are abelian (i.e. taking values in the Cartan subalgebra).

Asymptotically, for the n-th dyon, we may write [40]

$$\pm E_k^{(n)} = B_k^{(n)} = \tfrac{1}{2}\frac{x_k}{r^3}C^{(n)}\,, \tag{9.56}$$

where the $C^n = \mathrm{diag}[0,\ldots,1,-1,0,\ldots,0]$ (with 1 in the n-th position) for n from 1 to $N-1$, and $C^{(N)} = \mathrm{diag}[-1,0,\ldots,1]$. The dyon solutions were discovered by Bogolmolny et al. [41], and are known as BPS monopoles. A naive superposition of dyon fields is in general not a saddlepoint of the classical action; calorons, however, are not naive at all. A (topological) charge one caloron has N dyon constituents, and is indeed a solution of the Euclidean field equations.

A characteristic feature of the KvBLL caloron is the fact that the Polyakov loop holonomy (i.e. the Wilson line closed by periodicity in the time direction, before taking the trace) need not be a center element far away from the caloron. In general we can write the Polyakov loop holonomy as

$$\mathscr{P}(\mathbf{x}) = P \exp\left[i \int dt\, A_0(\mathbf{x}, t)\right]$$
$$= S\,\mathrm{diag}[e^{2\pi i\mu_1}, e^{2\pi i\mu_2},\ldots, e^{2\pi i\mu_N}]S^{-1}\,, \tag{9.57}$$

where S is some $SU(N)$ matrix which diagonalizes $\mathscr{P}$, and $\sum_n \mu_n = 0$. There are different caloron solutions, depending on the holonomy far from the caloron center. If $\mathscr{P}$ is a center element, then all the μ_n are equal; this is the "trivial" holonomy which characterizes the finite temperature instanton solutions found by Harrington and Shepard [42,43]. Caloron solutions allow for non-trivial holonomy. If we order the $\{\mu_n\}$ such that $\mu_i \leq \mu_{i+1}$, then the maximally "non-trivial" holonomy is where the $\{\mu_n\}$ are equally spaced and as far apart as possible, i.e.

$$\mu_n^{max} = -\frac{1}{2} - \frac{1}{2N} + \frac{n}{N}\,. \tag{9.58}$$

In this special case, $\mathrm{Tr}[\mathscr{P}] = 0$. In a vacuum dominated by calorons of this special type, a Polyakov loop at a given location would tend to fluctuate around $\mathrm{Tr}[\mathscr{P}] = 0$, with the probability density peaked at $\mathrm{Tr}[\mathscr{P}] = 0$. This is in sharp contrast to the center vortex confinement scenario, where the probability density would be peaked at center elements, and in fact Polykov loop holonomies are *only* center elements in center-projected configurations. The vanishing of the Polyakov loop expectation value in the vortex scenario is due to an unbroken center symmetry, leading to an exact cancellation between positive and negative contributions, whereas in a purely caloron configuration, with maximally non-trivial holonomy, the vanishing of the Polyakov loop is a property of the configuration itself, rather than arising from quantum fluctuations among such configurations.

One of the features of the KvBLL caloron is that the positions of dyon constituents are set by the parameters ("moduli") of the caloron solution. The constituents can be widely separated; in fact they can be placed anywhere, depending

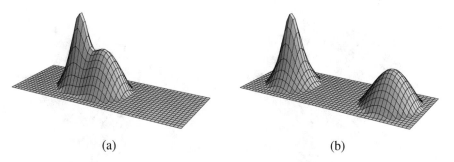

(a)   (b)

Fig. 9.8 Action density of an $SU(2)$ caloron in a timeslice $t = 0$, with $\mu_2 = -\mu_1 = 0.125$. The parameter ρ determines the separation of the constituent dyons; the figure on the left is at $\rho = 0.8$, and the figure on the right has $\rho = 1.6$. From Kraan and van Baal [44]

on the choice of moduli. An example for the $SU(2)$ group is shown in Fig. 9.8, which displays the action density of two $SU(2)$ caloron solutions in a timeslice at $t = 0$. Because the caloron is an instanton solution, one might think that the dyon constituents appear and then disappear in a certain time interval. In fact, this is true for small dyon separations. When the dyons are widely separated, however, they persist throughout the entire periodic time interval, as seen in Fig. 9.9.

The fact that calorons are saddlepoints of the Euclidean Yang-Mills action, and can have both non-trivial holonomy and widely separated monopole constituents, has led to the conjecture that these constituents form a kind of dyon gas, which gives rise to the necessary magnetic disorder for confinement. Then the Yang-Mills functional integral might be approximated by summing over dyon configurations.

In a remarkable paper, Diakonov and Petrov [40] derived analytically a confining quark-antiquark potential from Polyakov line correlators, and an area law for spacelike Wilson loops, from dyon-antidyon configurations in $D = 4$ dimensions, and showed that the string tension was the same in the two cases. These dyon configurations should dominate the vacuum at large scales if confinement can be traced to KvBLL calorons with maximally non-trivial holonomy (hereafter just "calorons"). The statistical weight of each dyon configuration is given by a certain determinant whose details will not concern us here.[5]

The abelian field strength is, in this case, controlled by variables $v_m(x)$, which appear in the partition function for the dyon ensemble. For SU(N) gauge theory this partition function has the form

$$Z = \int D\chi^\dagger \, D\chi \, Dv \, Dw \, \exp \int d^3x \left\{ \frac{T}{4\pi} \left(\partial_i \chi_m^\dagger \partial_i \chi_m + \partial_i v_m \partial_i w_m \right) \right.$$
$$\left. + f \left[(-4\pi \mu_m + v_m) \frac{\partial \mathcal{F}}{\partial w_m} + \chi_m^\dagger \frac{\partial^2 \mathcal{F}}{\partial w_m \partial w_n} \chi_n \right] \right\} , \tag{9.59}$$

---

[5]Except to note in passing the critical comments in [46].

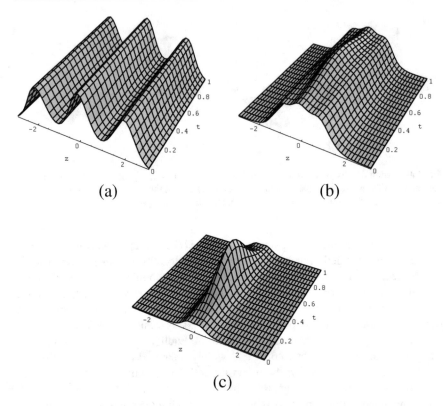

Fig. 9.9 Action density of $SU(3)$ calorons as a function of time and one space coordinate. Each caloron has three dyon constituents, and the dyon separations are increasing from right to left. Note the persistence in time of the widely separated consituent dyons. From Diakonov [45]

where the subscripts ($m = 1, .., N$) label the dyon type. For an explanation of the terms in this expression, see [40]. The abelian magnetic field $B_i = \frac{1}{2}\varepsilon_{ijk} f_{jk}$ due to the m-th dyon type is given by

$$[B_i(\mathbf{x})]_m = -\frac{T}{2}\partial_i v_m(\mathbf{x}) , \qquad (9.60)$$

where T is temperature. Note that this expression for B_i does not include Dirac strings, which have no effect on Wilson loops, but which are important in showing that $\nabla \cdot \mathbf{B} = 0$. Diakonov and Petrov were able to find saddlepoint solutions of the effective action with a spacelike Wilson loop as external source. These solutions generalize the solitonic solution found by Polyakov for compact QED$_3$, representing a monopole-antimonopole sheet along the minimal surface of the loop. The analysis provides a demonstration of the area-law falloff of a spacelike Wilson loop in the dyon ensemble, and an explicit calculation of the string tension in group representations of N-ality k. These authors also derived, from their dyon model, the following results:

1. At finite but low temperatures, the free energy of the dyon gas is minimized at the maximally non-trivial Polyakov holonomy, where $\text{Tr}[\mathscr{P}] = 0$.
2. k-string tensions extracted from Polyakov line correlators, and from spacelike Wilson loops, agree with one another, and follow the Sine Law. The calculation of spacelike string tensions is similar to Polyakov's determination of the string tension in compact QED_3.
3. At high temperatures, the free energy is minimized at the trivial holonomy (deconfinement), and there is a phase transition between the confined and deconfined phases at a critical temperature T_c, with $T_c/\sqrt{\sigma}$ in impressive agreement with lattice simulations.
4. $\text{Tr}[\mathscr{P}] = 0$ is also obtained in G_2 gauge theory, leading to a claim [45] that center symmetry is of no fundamental importance to confinement.

The Diakonov-Petrov treatment of the Yang-Mills vacuum as a dyon liquid of some kind has been further developed in a series of articles by Liu et al. [47–49]. The Diakonov-Petrov work was based on the assumption of Coulombic interactions among the dyons, which is a good approximation at they large dyon separations typical of a dilute gas. An improved treatment of dyon interactions by Larsen and Shuryak [50] takes account of the repulsive interactions due to the finite size dyon cores, and Debye screening of the long range Coulombic interaction. This improved treatment was incorporated in the work of [47–49]. In these references the dyon model was solved via mean field methods. The model was also simulated numerically in [51–53]. One of the main results coming out of this work was the observation, for SU(2) gauge theory with two flavors of light fundamental quarks, that the deconfinement and chiral symmetry restoration transitions occur at about the same dyon density.

Although the dyon model of the Yang-Mills vacuum is quite successful in many ways, there is still a question of whether such configurations actually exist in the vacuum state; one would like to see evidence that such configurations are found in vacuum configurations generated by lattice Monte Carlo simulations. This issue was first addressed by Gattringer in [54], with subsequent studies along the same lines by Bornyakov et al. [55], and Larsen et al. [56]. The idea is to focus on configurations generated from SU(3) lattice Monte Carlo simulations with topological charge $Q = \pm 1$, and to exploit two facts about dyons. First, assuming that there is a KvBLL caloron in such configurations, the zero mode of the lattice Dirac operator will be concentrated at the position of one of the three dyons composing the caloron. Secondly, the dyon chosen by the zero mode depends on the boundary conditions of the lattice Dirac operator in the temporal direction. This means that if the position of the zero mode in a given lattice configuration is sensitive to boundary conditions imposed on the lattice Dirac operator, i.e. if that position is seen to change upon changing boundary conditions then that is evidence of an underlying caloron configuration composed of well-separated dyons. One considers boundary conditions of the form

$$\psi(\mathbf{x}, t + \beta) = e^{i\phi} \psi(\mathbf{x}, t) \tag{9.61}$$

where $\beta = 1/T$ is the extension of the lattice in the Euclidean time direction. The special values $\phi = 0, \pi$ correspond to periodic and antiperiodic boundary conditions respectively. It should be understood that these boundary conditions are imposed on the lattice Dirac operator; there does not have to be dynamical fermions in the system. With μ_1, μ_2, μ_3 defined as in (9.57), ordered such that $\mu_1 \leq \mu_2 \leq \mu_3$, and defining also $\mu_4 = 1 + \mu_1$, it was shown in [57, 58] that the zero mode of the lattice Dirac operator will be concentrated on the m-th dyon for $2\pi\phi$ in the range $[\mu_m, \mu_{m+1}]$.

Let $\psi(x)$ be the zero mode of the lattice Dirac operator, and define the scalar density $\rho(x) = |\psi(x)|^2$, where summation over both color and Dirac indices is assumed. Figure 9.10 is a plot (from [54]) of the scalar density in the $y - z$ plane at fixed x and t coordinates, for a particular gauge field configuration in pure SU(3) gauge theory, generated by lattice Monte Carlo, at a temperature below the deconfinement transition, in the confined phase. The left hand plot is the scalar density with anti-periodic boundary conditions in the time direction, while the scalar density in the right hand plot is computed in the same lattice configuration, only with periodic boundary conditions imposed on the lattice Dirac operator. The change in the position of the peak, upon changing boundary conditions in the time direction, is consistent with the view that the $|Q| = 1$ topological charge is due to a caloron configuration which is composed of well-separated dyons.

A much more recent plot along the same lines due to Larsen et al. [56], is shown in Fig. 9.11. These authors have worked with $|Q| = 1$ lattice configurations on a $32^3 \times 8$ lattice which include the effects of dynamical fermions. The red, blue, green colors correspond to periodicity conditions (9.61) with $\phi = \pi, \pi/3, -\pi/3$ respectively, imposed, in each panel, on the same configuration. In this case the scalar density is displayed in the $x - y$ plane, at fixed z, after averaging over the time direction. The left panel has been obtained for a configuration at the critical

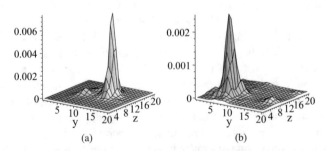

Fig. 9.10 Squared modulus of the zero-mode of the Dirac operator for anti-periodic temporal b.c. (l.h.s.) and periodic temporal b.c. (r.h.s.). We show y-z slices for a configuration in the confined phase, below the deconfinement temperature T_c. From Gattringer [54]

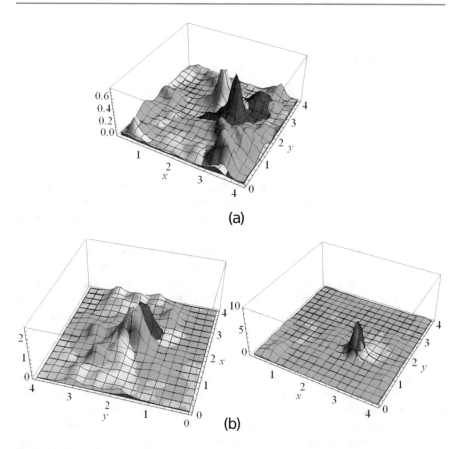

Fig. 9.11 Squared modulus of the fermion zero-mode of the overlap Dirac operator in the $x - y$ plane, summed over the temporal direction of three statistically independent QCD configurations (**a**) at T_c and (**b**) at $T = 1.08\ T_c$. The three different colors represent the zero-modes at temporal periodicity phases $\phi = \pi$ (red), $\phi = \pi/3$ (blue), $\phi = -\pi/3$ (green) respectively. Peak heights have been normalized corresponding to the height for $\phi = \pi$. The x and y coordinates are in units of $1/T$. From Larsen et al. [56]

temperature T_c, and just above the critical temperature at $T = 1.08 T_c$ in the middle and right panels. Once again, the message is that the zero mode is localized in different regions, which depend on the choice of boundary conditions, at least in the left and middle panels. In the right panel, the zero mode seems to be localized in the same region for all three boundary conditions, but Larsen et al. make the case that this configuration also represents a caloron with non-trivial holonomy, and that the spacetime profiles of the scalar densities fit the analytical expressions which are expected for these quantities in the dyon picture [56].

One issue that could be mentioned here is the probability distribution for Polyakov holonomies. If the Polyakov loop vanishes for a loop in the fundamental representation, then the corresponding Polyakov loop in the adjoint representation is negative. In fact, if $\mathscr{P}(\mathbf{x})$ is the Polyakov holonomy, one can express the probability density $\rho(g)$ that $\mathscr{P}(\mathbf{x})$ is equal to group element g via a character expansion. For, e.g., the case of $SU(2)$, we have

$$
\begin{aligned}
\rho[g] &= \langle \delta[\mathscr{P}(\mathbf{x}) - g] \rangle \\
&= \langle \sum_j \chi_j[\mathscr{P}(\mathbf{x})] \chi_j[g] \rangle \\
&\approx 1 + a_1 \chi_1[g] ,
\end{aligned}
\qquad (9.62)
$$

where $a_1 = \langle \chi_1[\mathscr{P}(\mathbf{x})] \rangle$ is the expectation value of the Polyakov line in the adjoint $(j = 1)$ representation. The $j =$ half-integer representations can be dropped, since they should vanish if center symmetry is unbroken, and the expectation values of $\mathscr{P}(\mathbf{x})$ in representations $j > 1$ are negligible. If $a_1 < 0$ then the maximally non-trivial holonomy is the most probable, in conformity with the dyon gas idea. If, on the other hand, $a_1 > 0$, then $\rho[\mathscr{P}]$ is peaked at $\mathscr{P} =$ a center element, as would be expected in the center vortex picture. The expectation value of an adjoint Polyakov line in the confined phase is actually quite small and difficult to measure, nevertheless there are measurements of this quantity in $SU(3)$ gauge theory just below the deconfinement transition, and the result is that the expectation value is small but positive [59].

Finally, if it were really true that the Polyakov line expectation value vanishes *exactly* in G_2 gauge theory at low temperatures, then center symmetry is indeed irrelevant to confinement, as claimed in [45]. However, it is much more likely that this prediction of the dyon gas approximation is just incorrect. In G_2 it is necessary to bind at least three gluons to screen a Polyakov line in the fundamental representation, and a gluelump of this sort will be very heavy. As a result, the Polyakov line in G_2 gauge theory will be very small indeed, and it will be exceedingly difficult to measure this tiny non-zero value by numerical simulation. But unless current ideas about charge screening are completely wrong, the Polyakov line in G_2 must be non-zero. For the same reason, any large G_2 Wilson loop can be screened by gluons, and the asymptotic string tension must vanish. This means that the lack of center symmetry implies the absence, at sufficiently large scales, of magnetic disorder. The fact that the dyon gas approximation misses this fact should not be taken as implying that center symmetry is irrelevant; instead it means that the dyon gas approximation is itself missing something which is important to the asymptotic physics.

9.6   Critique: N-ality and Multiple Winding Loops

All of the confinement mechanisms discussed in this chapter have something in common: there is some choice of gauge in which the large scale quantum fluctuations responsible for disordering Wilson loops are essentially abelian, and are found primarily in the gauge fields associated with the Cartan subalgebra of the gauge group. In a caloron ensemble, for example, while the dyon cores may be essentially non-abelian, there exists a gauge in which the long range field which diverges from the dyon cores, and which is responsible for confinement in this picture, lies entirely in the Cartan subalgebra. For the SU(2) gauge group, which is sufficient for our purposes, let this abelian field be the A_μ^3 color component. Then if all we are interested in is the area law falloff and corresponding string tension extracted from large Wilson loops, and not in perimeter law or short-range contributions from small Wilson loops, we can make the "abelian dominance" approximation

$$
\begin{aligned}
W(C) &= \tfrac{1}{2}\langle \mathrm{Tr} P \exp\left[i \oint_C dx^\mu A_\mu^a \frac{\sigma^a}{2}\right]\rangle \\
&\approx \tfrac{1}{2}\langle \mathrm{Tr} \exp\left[i \oint_C dx^\mu A_\mu^3 \frac{\sigma^3}{2}\right]\rangle \\
&= \langle \exp\left[i\tfrac{1}{2} \oint_C dx^\mu A_\mu^3\right]\rangle \\
&= \langle \exp\left[i\tfrac{1}{2} \int_S d\sigma^{\mu\nu} f_{\mu\nu}^3\right]\rangle ,
\end{aligned}
\tag{9.63}
$$

where $f_{\mu\nu}^3$ is the corresponding abelian field strength. Equivalently, we could imagine integrating out all other degrees of freedom to arrive at an effective abelian theory, i.e.

$$
e^{-S_{ab}[A_\mu^3]} = \int DA_\mu^1 DA_\mu^2 D(\text{ghosts}) \, e^{(-S+S_{gf})}
\tag{9.64}
$$

where S_{gf} is the gauge-fixing part of the action. Then the abelian dominance approximation is equivalently

$$
W(C) \approx \frac{1}{Z} \int DA_\mu^3 \exp\left[i\tfrac{1}{2} \oint_C dx^\mu A_\mu^3\right] e^{-S_{ab}[A_\mu^3]}
\tag{9.65}
$$

In a numerical simulation, an expectation value of some observable is the average taken over a very large number of samples drawn from some probability distribution, in this case $\exp[-S_{ab}[A_\mu^3]]$. The question we are concerned with is: what do typical configurations drawn from the abelian field distribution look like? Do they resemble what is predicted by monopole plasma, caloron gas, and dual superconductor models?

Monopole, dyon, and dual superconductor models all make definite predictions about the distribution of the abelian magnetic field strength on large scales, and how these abelian fluctuations generate an area law falloff for Wilson loops. But there are many types of Wilson loops; the simple planar contours that are generally considered is only one small subclass. If the abelian mechanisms are correct, and abelian vacuum fluctuations have the large-scale character that they specify, then there are other types of Wilson loop observables that should be considered. In this critique section we will point to the relevance of multiple winding loops, i.e. loops which wind around a given contour more than once, as an important test of the models discussed in this chapter.

9.6.1   Area Law Falloff for Double-Winding Loop

We will continue to treat the SU(2) case for simplicity, but the generalization to SU(N) theories is straightforward. Let C_1 and C_2 be two co-planar loops, with C_1 lying entirely in the minimal area of C_2, which share a point $\mathbf{x}$ in common. Consider a Wilson loop in SU(2) gauge theory which winds once around C_1 and once, winding with the same orientation, around C_2, as indicated in Fig. 9.12. It will also be useful to consider Wilson loop contours in which C_1 lies mainly in a plane displaced in a transverse direction from the plane of C_2 by a distance δz comparable to a correlation length in the gauge theory. Such a contour is indicated in Fig. 9.13. We will refer to both of these cases as "double-winding" Wilson loops. In both cases we imagine that the extension of loops C_1, C_2 is much larger than a correlation length, so in the latter example the displacement of loop C_1 from the plane of C_2 is small compared to the size of the loops. Let A_1, A_2 be the minimal areas of loops C_1, C_2 respectively. What predictions can be made about the expectation value $W(C)$ of a double-winding Wilson loop, as a function of areas A_1 and A_2?

The point that we wish to show here is that the monopole-dyon-dual supercon-ductor models predict a "sum of areas" falloff for the contours in Fig. 9.13,

$$W(C) \sim \exp[-\sigma(A_1 + A_2) - \mu P]\,, \tag{9.66}$$

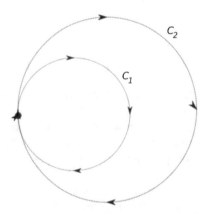

Fig. 9.12 A double winding loop, which runs once around contour C_1, and once around the coplanar loop C_2

Fig. 9.13 A "shifted" double winding loop, in which contours C_1 and C_2 lie in planes parallel to the $x - y$ plane, but are displaced from one another in the transverse direction by distance δz, and are connected by lines running parallel to the z-axis

where A_1, A_2 is the minimal area enclosed by C_1, C_2 respectively, and P is the sum of the lengths of C_1 and C_2. In the center vortex the leading behavior is governed by a "difference in areas" expression.

$$W(C) \sim \exp[-\sigma|A_2 - A_1|] . \tag{9.67}$$

Let us begin with the sum-of-areas law. We assume that the large scale fluctuations are abelian in character, so that (9.63) holds, and the distribution of abelian field strength is as predicted by the monopole-dyon-dual superconductor models discussed. Then

$$
\begin{aligned}
W(C) &= \tfrac{1}{2}\left\langle \mathrm{Tr} P \exp\left[i \oint_C dx^\mu A_\mu^a \frac{\sigma^a}{2}\right]\right\rangle \\
&\approx \left\langle \exp\left[i\tfrac{1}{2} \oint_C dx^\mu A_\mu^3\right]\right\rangle \\
&= \left\langle \exp\left[i\tfrac{1}{2} \oint_{C_1} dx^\mu A_\mu^3\right] \exp\left[i\tfrac{1}{2} \oint_{C_2} dx^\mu A_\mu^3\right]\right\rangle .
\end{aligned}
$$
$$\tag{9.68}$$

If loops C_1 and C_2 are sufficiently far apart, then the expectation value of the product is approximately the product of the expectation values, i.e.

$$
\begin{aligned}
W(C) &\approx 2\left\langle \exp\left[i\tfrac{1}{2} \oint_{C_1} dx^\mu A_\mu^3\right]\right\rangle\left\langle \exp\left[i\tfrac{1}{2} \oint_{C_2} dx^\mu A_\mu^3\right]\right\rangle \\
&\approx \exp[-\sigma(A_1 + A_2)] ,
\end{aligned}
$$
$$\tag{9.69}$$

which is the sum-of-areas falloff. In a monopole plasma, the double winding loop can be interpreted as inserting two independent current loops into the plasma. Monopoles (or monopole currents) will respond by forming a monopole-antimonopole layer at the minimal surface of each loop. The argument in the case of the dual superconductor is similar; we imagine that loops C_1 and C_2 are rectangular and oriented parallel to the $x - t$ plane, but displaced along the z-axis. In a time slice, this setup represents a pair of positive charges, a distance δz apart, interacting

Fig. 9.14 A time slice of shifted rectangular timelike loops can be interpreted as representing two static particles on one side, and two static antiparticles on the other. In the dual abelian Higgs model, the pairs of $\pm$ charges are connected by a pair of electric flux tubes

with a pair of negative charges, also a distance δz apart, and two electric flux form, as seen in Fig. 9.14. The energy is $\sigma(L_1 + L_2)$, where L_1, L_2 are the lengths of the two flux tubes, and this implies, from the usual relationship between Wilson loops and static potentials, a sum-of-areas falloff for the Wilson loop.

Now, what happens as $\delta z \to 0$? We would argue that this limit does not really change the sum-of-areas behavior. For a dual abelian Higgs model with couplings corresponding to a Type II (dual) superconductor, electric flux tubes tend to repel. So as the two positive and two negative charges converge, we would still expect to find two electric flux tubes separated by roughly the vortex width, and the sum-of-areas rule does not change qualitatively. It has also been suggested [60] that the relevant dual abelian Higgs model is weakly Type I, near the crossover from Type I to Type II behavior. In a Type I dual superconductor the electric flux tubes would attract, and presumably merge. The energy per unit length of the merged flux tubes would then be somewhat less than the sum of energies per unit length of two flux tubes of minimal electric flux. The double-winding Wilson loop falloff would then be a little less than sum-of-areas, but this slight difference would not affect our argument in any essential way.

In the case of a $D = 3$ monopole Coulomb gas we can be a little more explicit, following closely the old arguments of Ref. [8]. We begin with shifted loops, both oriented parallel to the $x - y$ plane, with C_1 at $z = 0$ and C_2 at $z = \delta z$. Then, by the standard manipulations introduced by Polyakov, we have

$$
\langle W(C) \rangle = \frac{1}{Z_{mon}} \int D\chi(r) \, \exp\left[-\frac{g^2}{4\pi} \int d^3r \left(\tfrac{1}{2}(\partial_\mu(\chi - \eta_{S(C)})^2 \right.\right.
$$
$$
\left.\left. -M^2 \cos \chi(r) \right) \right],
\tag{9.70}
$$

where

$$
-\partial^2 \eta_{S(C)} = 2\pi \delta'(z)\theta_{S_2}(x, y) + 2\pi \delta'(z - \delta z)\theta_{S_1}(x, y) ,
\tag{9.71}
$$

and $\theta_{S_{1(2)}}(x, y) = 1$ if x, y lie in the minimal area of C_1 (C_2), and is zero otherwise. Assuming $\delta z \gg 1/M$, an approximate saddlepoint solution is the superposition

$$
\begin{aligned}
\chi = \text{sign}\, z \cdot 4 \arctan(e^{-M\,|z|}) \theta_{S_2}(x, y) \\
+\text{sign}\,(z - \delta z) \cdot 4 \arctan(e^{-M\,|z-\delta z|}) \theta_{S_1}(x, y) \,.
\end{aligned}
\tag{9.72}
$$

As $\delta z \to 0$ we may still choose the surfaces S_1, S_2 to be displaced from one another in the z-direction, except near the loop boundaries. If we take this displacement to be $d \gg 1/M$, then (9.72) with $\delta z \to d$ is still an approximate solution for large loops, where the areas of S_1, S_2 are still nearly minimal, and nearly parallel to the $x - y$ plane. In either case we have two monopole-antimonopole sheets where the x, y coordinates of S_1, S_2 coincide, and one sheet where x, y lies in S_2, but not in S_1. The result is a sum-of-areas falloff for the double-winding Wilson loop. However, at $\delta z = 0$ there is another approximate solution, with discontinuities only on the minimal areas of C_1 and C_2, that was found in [8]. For x, $y \in$ the minimal area of C_1, and $d \gg 1/M$ but small compared to the extension of the loop, the solution is

$$
\begin{aligned}
\chi = \theta(z)4 \arctan(e^{-M(z-d)}) \\
+\theta(-z)[4 \arctan(e^{-M(z+d)}) - 2\pi] \,,
\end{aligned}
\tag{9.73}
$$

while for x, $y \in$ the minimal area between C_1 and C_2, the solution is the standard Polyakov soliton for a single-winding loop

$$
\chi = \text{sign}\, z \cdot 4 \arctan(e^{-M\,|z|}) \,.
\tag{9.74}
$$

In both cases x, y are far from the loop perimeters. The result is again a sum-of-areas falloff.

For a monopole plasma in $D = 4$ dimensions, we can use the fact that in the confined phase this model can be mapped into compact QED at strong couplings [61]. It is trivial to calculate the double-winding Wilson loop in compact QED$_4$ at strong lattice couplings, and the result is essentially a sum-of-areas falloff.

The Diakonov-Petrov calculation of spacelike Wilson loops in $D = 3 + 1$ dimensions is, as already mentioned, a generalization of the Polyakov calculation in $D = 3$ dimensions. As in the Polyakov calculation, the analytical solution involves a soliton peaked at the minimal area of the spacelike loop, and which falls to zero in either direction transverse to the loop. The sum-of-areas result follows fairly trivially for the shifted double-winding loop so long as δz is greater than the thickness of this soliton.

In contrast, in the center vortex picture of confinement, and also in strong coupling lattice gauge theory, large double-winding loops obey the difference of areas law (9.67). The argument goes as follows. It is assumed that the loops are so large that the thickness of center vortices can be neglected. For co-planar loops, if a vortex pierces the minimal area of loop A_1, it will multiply the holonomy around loop C_1 by -1, and also multiply the holonomy around C_2 by -1, producing no effect whatever on the double-winding loop (unless the vortex crosses a loop perimeter, which can only result in a perimeter-law contribution). So the vortex crossing can only produce an effect if it pierces the minimal area of C_2 but not the minimal area of C_1 (difference of areas $A_2 - A_1$). This supplies an overall factor of -1 to the double-winding holonomy. By the usual argument (see section 6.1) this results in a "difference-of-areas" falloff (9.67). A slight shift of loop C_1 by δz in the transverse direction does not make any difference to the argument, providing the scales of A_1 and A_2 are so large compared to δz that a vortex piercing the smaller area A_1 is guaranteed to also pierce the larger area A_2.

The double-winding loop is also easily computed in strong-coupling SU(2) lattice gauge theory, with the result

$$W(C) = -\tfrac{1}{2} \exp[-\sigma |A_2 - A_1|]$$

$$\sigma = -\log\left[\frac{I_2(\beta)}{I_1(\beta)}\right] , \qquad (9.75)$$

which is again a difference-of-areas law. A small shift δz in the loop C_1 will not affect this answer. The center vortex model does not pick up the same overall sign, but a model which only considers center vortex contributions to large Wilson loops is not complete enough to pick up either the perimeter law behavior or any overall constant, but only the area-law falloff.

The question is then: which is right, sum-of-areas or difference-of-areas? The question has been investigated in [62] by lattice Monte Carlo simulations, and the answer is unequivocally that the double-winding loops follow a difference-in-areas law; a sample of the numerical evidence is shown in Fig. 9.15. From this it seems that the monopole-dyon-dual superconductor pictures, as they stand, are incompatible with the numerical results. Assuming that solitonic objects such as magnetic monopoles are important in the Yang-Mills vacuum, a possible way out of

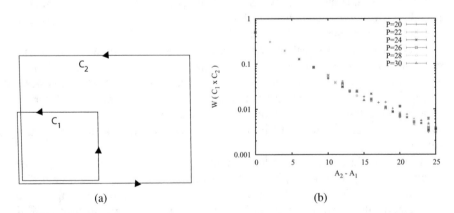

(a)   (b)

Fig. 9.15 Wilson loop expectation values $W(C_1 \times C_2)$ at fixed perimeter P vs. difference in area (**b**), for the rectangular contours shown in (**a**). Two sides of loops C_1 and C_2 overlap on the lattice, although they are drawn as slightly displaced. From Greensite and Hollwieser [62]

Fig. 9.16 For the same situation depicted in Fig. 9.14, insertion of a positively and negatively charged W boson neutralizes the widely separated positive and negative charges. Then there are only flux tubes between the positive static charges and the W^{--}, and (separately) the negative static charges and the W^{++}, leading to a difference-in-areas law

the conundrum would be if their abelian flux were somehow collimated into center vortex configurations, as already outlined in Sect. 9.4.1. This would be the true distribution of abelian flux governed by the abelian effective action S_{ab}.

In the abelian effective action S_{ab}, the effect of gauge degrees freedom outside the Cartan subalgebra, i.e. the W-bosons, have already been taken into account. It is not hard to see that the W-bosons must be responsible for a difference-of-areas law. If we return to Fig. 9.14, we see that the objects on the left are collectively double-charged positive, and the objects on the right are collectively double-charged negative. But this means that the objects on the left, and the objects on the right, can bind to W-bosons, which effectively neutralize them, as illustrated in Fig. 9.16. While this points to the fact that the difference-of-areas law is associated with non-abelian degrees of freedom it should be understood that the distribution of abelian flux in the vacuum is governed by S_{ab}, and in this action there are no longer any W-boson degrees of freedom which could shield a double-winding Wilson loop and give rise to the difference of area law. Only a particular type of vacuum fluctuation, i.e. center vortices, can have that effect. After integrating out the W-boson degrees of freedom to obtain S_{ab}, the vacuum fluctuations of S_{ab} must have the vortex character we have described.

As a simple illustration of how double-charged matter fields can change a monopole Coulomb gas into a system of vortices, we consider an abelian Higgs model with a double-charged matter field

$$Z = \int D\rho D\theta_\mu \, \exp\left[\beta \sum_p \cos(\theta(p)) + \tfrac{1}{2}\lambda \sum_{x,\mu} \left\{\rho^*(x)e^{2i\theta_\mu(x)}\rho(x+\widehat{\mu}) + \text{c.c}\right\}\right],$$

(9.76)

with $\beta \ll 1$ (confinement) and $\lambda \gg 1$. In this case, rewriting the theory in monopole variables actually obscures the underlying physics. The confining field configurations are no longer Coulombic fields emanating from monopole charges. Rather, the confining configurations are thin Z_2 vortices—a fact which is invisible in the monopole formulation. To see this, go to the unitary gauge $\rho = 1$, which preserves a residual Z_2 gauge invariance, and make the field decomposition

$$\exp[i\theta_\mu(x)] = z_\mu(x) \exp[i\widetilde{\theta}_\mu(x)],$$

(9.77)

where

$$z_\mu(x) \equiv \text{sign}[\cos(\theta_\mu(x))] , \qquad (9.78)$$

and

$$Z = \prod_{x,\mu} \sum_{z_\mu(x)=\pm1} \int_{-\pi/2}^{\pi/2} \frac{d\widetilde{\theta}_\mu(x)}{2\pi}$$

$$\exp\left[\beta \sum_p Z(p) \cos(\widetilde{\theta}(p)) + \lambda \sum_{x,\mu} \cos(2\widetilde{\theta}_\mu(x)) \right] . \qquad (9.79)$$

This decomposition separates lattice configurations into Z_2 vortex degrees of freedom (the $z_\mu(x)$), and small non-confining fluctuations around these vortex configurations, strongly peaked at $\widetilde{\theta} = 0$. One can easily show, for $\beta \ll 1$, $\lambda \gg 1$, that

$$\left\langle \exp[in\theta(C)] \right\rangle \approx \left\langle Z^n(C) \right\rangle \left\langle \exp[in\widetilde{\theta}(C)] \right\rangle , \qquad (9.80)$$

with

$$\left\langle Z^n(C) \right\rangle = \begin{cases} \exp[-\sigma A(C)] & n \text{ odd} \\ 1 & n \text{ even} \end{cases}$$

$$\left\langle \exp[in\widetilde{\theta}(C)] \right\rangle = \exp[-\mu n^2 P(C)] , \qquad (9.81)$$

where $Z(C)$ is the product of $z_\mu(x)$ link variables around the loop C. This establishes that the confining fluctuations, in this coupling range, are entirely due to thin vortices identified by the decomposition (9.77) in unitary gauge. It is clear that the addition of a charge-2 matter field has resulted in a qualitative change in the physics of confinement. Yet the transition from a monopole Coulomb gas mechanism to a vortex dominance mechanism is essentially invisible if the gauge+matter theory is rewritten in terms of monopole + electric current variables, which in this case tend to obscure, rather than illuminate, the nature of the confining fluctuations.

References

1. S. Coleman, The magnetic monopole fifty years later, in *The Unity of the Fundamental Interactions*, ed. by A. Zichichi (Kluwer Academic Publishers, Dordrecht, 1983)
2. G. 't Hooft, Gauge fields with unified weak, electromagnetic, and strong interactions, in *High-Energy Physics: Proceedings of the EPS International Conference*, Palermo, 1975, ed. by A. Zichichi

3. S. Mandelstam, Vortices and quark confinement in non-Abelian gauge theories. Phys. Reports **23C**, 245 (1976)
4. A. Polyakov, *Gauge Fields and Strings* (Harwood Academic Publishers, 1987) ; Compact gauge fields and the infrared catastrophe. Phys. Lett. B **59**, 82 (1975)
5. T. DeGrand, D. Toussaint, Topological excitations and Monte Carlo simulation of Abelian gauge theory. Phys. Rev. D **22**, 2478 (1980)
6. A. Polyakov, Quark confinement and topology of gauge groups. Nucl. Phys. B **120**, 429 (1977)
7. T. Banks, R. Myerson, J.B. Kogut, Phase transitions in Abelian lattice gauge theories. Nucl. Phys. B **129**, 493 (1977)
8. J. Ambjørn, J. Greensite, Center disorder in the 3D Georgi-Glashow model. J. High Energy Phys. **05**, 004 (1998). arXiv: hep-lat/9804022
9. G.N. Watson, Three triple integrals. Quart. J. Math. **10**, 266 (1939)
10. M. Gopfert, G. Mack, Proof of Confinement of Static Quarks in Three-Dimensional U(1) Lattice Gauge Theory for All Values of the Coupling Constant. Commun. Math. Phys. **82** (1981) 545
11. G. 't Hooft, Magnetic monopoles in unified gauge theories. Nucl. Phys. **B79**, 276 (1974)
12. A. M. Polyakov, "Particle spectrum in the quantum field theory," JETP Letters **20**, 194 (1974)
13. V.G. Bornyakov, E.M. Ilgenfritz, V.K. Mitrjushkin, A.M. Zadorozhnyi, M. Muller-Preussker, Investigation of the vacuum structure of the georgi-glashow model on the lattice. Z. Phys. C **42**, 633 (1989)
14. G. Schierholz, J. Seixas, M. Teper, Patterns of Symmetry Restoration in Gauge - Scalar Theories: A Monte Carlo Simulation of the SO(3) Georgi-Glashow Model. Phys. Lett. B **157**, 209 (1985)
15. R.C. Brower, D.A. Kessler, T. Schalk, H. Levine, M. Nauenberg, The SU(2) adjoint Higgs model. Phys. Rev. D **25**, 3319 (1982)
16. M. Shifman, M. Ünsal, QCD-like theories on $R_3 \times S_1$: a smooth journey from small to large $r(S_1)$ with double-trace deformations. Phys. Rev. D **78**, 065004 (2008). arXiv:0802.1232 [hep-th]
17. M. Ünsal, Magnetic bion condensation: a new mechanism of confinement and mass gap in four dimensions. Phys. Rev. **D80**, 065001 (2009). arXiv:0709.3269 [hep-th]
18. H.B. Nielsen, P. Olesen, Vortex line models for dual strings. Nucl. Phys. **B61**, 45 (1973)
19. N. Seiberg, E. Witten, Monopole condensation, and confinement in $N = 2$ supersymmetric Yang-Mills theory. Nucl. Phys. B **426**, 19 (1994). Erratum-ibid. B **430**, 485 (1994). arXiv:hep-th/9407087
20. Monopoles, duality and chiral symmetry breaking in $N = 2$ supersymmetric QCD. Nucl. Phys. B **431**, 484 (1994). arXiv:hep-th/9408099
21. A. Bilal, Duality in $N = 2$ SUSY SU(2) Yang-Mills Theory: A pedagogical introduction to the work of Seiberg and Witten. arXiv:hep-th/9601007
22. L. Alvarez-Gaume, S.F. Hassan, Introduction to S-duality in N = 2 supersymmetric gauge theories: a pedagogical review of the work of Seiberg and Witten. Fortsch. Phys. **45**, 159 (1997). arXiv:hep-th/9701069
23. W. Lerche, Introduction to Seiberg-Witten theory and its stringy origin. Nucl. Phys. Proc. Suppl. **55B**, 83 (1997). Fortsch. Phys. **45**, 293 (1997). arXiv:hep-th/9611190
24. M.R. Douglas, S.H. Shenker, Dynamics of supersymmetric SU(N) gauge theory. Nucl. Phys. **B447**, 271 (1995)
25. G. 't Hooft, Topology of the gauge condition and new confinement phases in non-Abelian gauge theories. Nucl. Phys. B **190** [FS3], 455 (1981)
26. A. Kronfeld, M. Laursen, G. Schierholz, U.-J. Wiese, Monopole condensation and color confinement. Phys. Lett. B **198**, 516 (1987)
27. G.S. Bali, V. Bornyakov, M. Muller-Preussker, K. Schilling, Dual superconductor scenario of confinement: a systematic study of Gribov copy effects. Phys. Rev. D **54**, 2863 (1996). arXiv:hep-lat/9603012
28. H. Shiba, T. Suzuki, Monopoles and string tension in SU(2) QCD. Phys. Lett. B **333**, 461 (1994). hep-lat/9404015

29. J. Stack, S. Neiman, R. Wensley, String tension from monoples in SU(2) lattice gauge theory. Phys. Rev. D **50**, 3399 (1994) [hep-lat/9404014]
30. L. Del Debbio, A. Di Giacomo, G. Paffuti, Phys. Lett. B **349**, 513 (1995). arXiv:hep-lat/9403013
31. A. Di Giacomo, B. Lucini, L. Montesi, G. Paffuti, Detecting superconductivity in the ground state of gauge theory. Phys. Rev. D **61**, 034503 (2000). arXiv:hep-lat/9906024
32. J. Greensite, B. Lucini, Is confinement a phase of broken dual gauge symmetry? Phys. Rev. D **78**, 085004 (2008). arXiv:0806.2117 [hep-lat]
33. J. Ambjørn, J. Giedt, J. Greensite, Vortex structure versus monopole dominance in Abelian projected gauge theory. J. High Energy Phys. **02**, 033 (2000). arXiv:hep-lat/9907021
34. T. Suzuki, I. Yotsuyanagi, A possible evidence for Abelian dominance in quark confinement. Phys. Rev. D **42**, 4257 (1990)
35. F.V. Gubarev, A.V. Kovalenko, M.I. Polikarpov, S.N. Syritsyn, V.I. Zakharov, Fine tuned vortices in lattice SU(2) gluodynamics. Phys. Lett. B **574**, 136 (2003). arXiv:hep-lat/0212003
36. A.A. Belavin, A.M. Polyakov, A.S. Shvarts, Yu.S. Tyupkin, Pseudoparticle solutions of the Yang-Mills equations. Phys. Lett. B **59**, 85 (1975)
37. T.C. Kraan, P. van Baal, Periodic instantons with non-trivial holonomy. Nucl. Phys. B **533**, 627 (1998). arXiv:hep-th/9805168
38. T.C. Kraan, P. van Baal, Exact T-duality between calorons and Taub - NUT spaces. Phys. Lett. B **428**, 268 (1998). arXiv:hep-th/9802049
39. K.M. Lee, C.H. Lu, SU(2) calorons and magnetic monopoles. Phys. Rev. D **58**, 025011 (1998). arXiv:hep-th/9802108
40. D. Diakonov, V. Petrov, Confining ensemble of dyons. Phys. Rev. D **76**, 056001 (2007). arXiv:0704.3181 [hep-th]
41. E.B. Bogomol'ny, Stability of classical solutions. Sov. J. Nucl. **24**, 449 (1976); M.K. Prasad, C.M. Sommerfield, An exact classical solution for the 't Hooft monopole and the Julia-Zee dyon. Phys. Rev. Lett. **35**, 760 (1975)
42. B.J. Harrington, H.K. Shepard, Thermodynamics of the Yang-Mills gas. Phys. Rev. D **18**, 2990 (1978)
43. B.J. Harrington, H.K. Shepard, Periodic Euclidean solutions and the finite temperature Yang-Mills gas. Phys. Rev. D **17**, 2122 (1978)
44. T.C. Kraan, P. van Baal, Constituent monopoles without gauge fixing. Nucl. Phys. Proc. Suppl. **73**, 554 (1999). arXiv:hep-lat/9808015
45. D. Diakonov, Topology and confinement. Nucl. Phys. Proc. Suppl. **195**, 5 (2009). arXiv:0906.2456 [hep-ph]
46. F. Bruckmann, S. Dinter, E.M. Ilgenfritz, M. Muller-Preussker, M. Wagner, Cautionary remarks on the moduli space metric for multi-dyon simulations. Phys. Rev. D **79**, 116007 (2009). arXiv:0903.3075 [hep-ph]
47. Y. Liu, E. Shuryak, I. Zahed, Confining dyon-antidyon Coulomb liquid model. I. Phys. Rev. D **92**(8), 085006 (2015). arXiv:1503.03058 [hep-ph]
48. Y. Liu, E. Shuryak, I. Zahed, Light quarks in the screened dyon-antidyon Coulomb liquid model. II. Phys. Rev. D **92**(8), 085007 (2015) arXiv:1503.09148 [hep-ph]
49. Y. Liu, E. Shuryak, I. Zahed, The instanton-dyon liquid model III: finite chemical potential. Phys. Rev. D **94**(10), 105011 (2016). arXiv:1606.07009 [hep-ph]
50. R. Larsen, E. Shuryak, Classical interactions of the instanton-dyons with antidyons. Nucl. Phys. A **950**, 110 (2016). arXiv:1408.6563 [hep-ph]
51. R. Larsen, E. Shuryak, Interacting ensemble of the instanton-dyons and the deconfinement phase transition in the SU(2) gauge theory. Phys. Rev. D **92**(9), 094022 (2015). arXiv:1504.03341 [hep-ph]
52. R. Larsen, E. Shuryak, Instanton-dyon ensemble with two dynamical quarks: the chiral symmetry breaking. Phys. Rev. D **93**(5), 054029 (2016). arXiv:1511.02237 [hep-ph]
53. R. Larsen, E. Shuryak, Instanton-dyon ensembles with quarks with modified boundary conditions. Phys. Rev. D **94**(9), 094009 (2016). arXiv:1605.07474 [hep-ph]

54. C. Gattringer, Calorons, instantons and constituent monopoles in SU(3) lattice gauge theory. Phys. Rev. D **67**, 034507 (2003). [hep-lat/0210001]

55. V.G. Bornyakov, E.-M. Ilgenfritz, B.V. Martemyanov, M. Muller-Preussker, Dyons near the transition temperature in lattice QCD. Phys. Rev. D **93**(7), 074508 (2016). arXiv:1512.03217 [hep-lat]

56. R.N. Larsen, S. Sharma, E. Shuryak, The topological objects near the chiral crossover transition in QCD. Phys. Lett. B **794**, 14 (2019). arXiv:1811.07914 [hep-lat]

57. M. Garcia Perez, A. Gonzalez-Arroyo, C. Pena, P. van Baal, Weyl-Dirac zero mode for calorons. Phys. Rev. D **60**, 031901 (1999) [hep-th/9905016]

58. M.N. Chernodub, T.C. Kraan, P. van Baal, Exact fermion zero mode for the new calorons. Nucl. Phys. Proc. Suppl. **83**, 556 (2000). [hep-lat/9907001]

59. S. Gupta, K. Huebner, O. Kaczmarek, Renormalized Polyakov loops in many representations. Phys. Rev. D **77**, 034503 (2008). arXiv:0711.2251 [hep-lat]

60. Y. Koma, M. Koma, E.M. Ilgenfritz, T. Suzuki, A detailed study of the Abelian projected SU(2) flux tube and its dual Ginzburg-Landau analysis. Phys. Rev. D **68**, 114504 (2003) [hep-lat/0308008]

61. J. Smit, A. van der Sijs, Monopoles and confinement. Nucl. Phys. B **355**, 603 (1991)

62. J. Greensite, R. Höllwieser, Double-winding Wilson loops and monopole confinement mechanisms. Phys. Rev. D **91**(5), 054509 (2015). arXiv:1411.5091 [hep-lat]

Coulomb Confinement

Abstract

The Gribov horizon, Neuberger's theorem, and the Gribov-Zwanziger confinement scenario. Coulomb confinement as a necessary condition for confinement. Numerical results for the Coulomb potential on the lattice, and the low-lying Faddeev-Popov eigenvalue spectrum. The role of center vortices in Coulomb confinement.

In Coulomb gauge, the longitudinal electric field $\mathbf{E}_L^a = -\nabla\phi^a$ associated with a static charge is obtained by solving the Gauss Law $\mathbf{D} \cdot \mathbf{E}^a = \rho^a$ for ϕ^a, and the energy density E_L^2 associated with this electric field, integrated over all space, is the Coulomb energy. There is an ongoing effort, pioneered originally by Gribov [1] and Zwanziger [2,3], to demonstrate that the Coulomb potential associated with quark-antiquark sources is confining. The argument that the Coulomb potential might have this property relies, in an essential way, on the fact that in Coulomb gauge the functional integration cannot range over all possible gauge fields satisfying the gauge condition, but instead must be restricted to a subspace in the space of all gauge fields, bounded by a hypersurface known as the *Gribov horizon*.

10.1 The Gribov Horizon

We have already seen that the Coulomb and Landau gauge conditions do not fix the gauge uniquely. In each case there is some remnant symmetry left, which may or may not be spontaneously broken, depending on the gauge theory and the gauge and matter coupling constants. Beyond this, however, there is a further ambiguity, first pointed out by Gribov [1]: In a gauge theory, each gauge field is a member of a set of gauge-equivalent configurations known as a *gauge orbit*. If we imagine that the gauge-fixing condition restricts the functional integral to a certain subspace

© Springer Nature Switzerland AG 2020

J. Greensite, *An Introduction to the Confinement Problem*, Lecture Notes in Physics 972, https://doi.org/10.1007/978-3-030-51563-8_10

$\mathscr{S}$ of the space of all gauge fields, then it turns out that a typical gauge orbit intersects this gauge-fixing hypersurface many times (not counting the multiplicity due to remnant symmetries), and these multiple intersections are known as *Gribov copies*. At each Gribov copy, the Faddeev-Popov (F-P) determinant may be positive or negative, which raises the alarming possibility that by summing over all copies, the functional integral may actually vanish! In fact, something of the sort is exactly what happens in BRST quantization of lattice gauge theories at the non-perturbative level, according to a famous result [4] known as

Neuberger's Theorem. *Let Q be a gauge or BRST invariant observable in a lattice gauge theory with compact link variables, with S_{gf} the gauge-fixing part of the action, and let $c, \bar{c}, b$ denote the ghost, antighost, and auxiliary fields, respectively, that arise in the BRST quantization procedure.[1] Then*

$$\langle Q \rangle = \frac{\int DU\,Dc\,D\bar{c}\,Db\ Qe^{-(S+S_{gf})}}{\int DU\,Dc\,D\bar{c}\,Db\ e^{-(S+S_{gf})}}$$

$$= \frac{0}{0} . \tag{10.1}$$

This means, of course, that BRST quantization is not really defined at the non-perturbative level. The problem can be traced, in the Faddeev-Popov approach, to the exact cancellation of contributions due to Gribov copies with F-P determinants of opposite sign [4].

It was suggested by Gribov that the domain of functional integration in Coulomb and Landau gauges should be restricted to a subspace, now known as the *Gribov region*, where the eigenvalues of the F-P operator, and hence the F-P determinant, are all positive. In this proposal the different Gribov copies cannot cancel. Consider, in the continuum, the gauge which minimizes the quantity

$$R[A] = \int d^4x \ \mathrm{Tr}[A_\mu(x)A_\mu(x)] . \tag{10.2}$$

If $A_\mu(x)$ is a (local) minimum of $R[A]$, and

$$^\theta A_\mu(x) = e^{i\theta^a(x)L_a} A_\mu(x) e^{-i\theta^a(x)L_a} - \frac{i}{g} e^{i\theta^a(x)L_a} \partial_\mu e^{-i\theta^a(x)L_a} \tag{10.3}$$

is a gauge transformation of $A_\mu(x)$ (with $\{L_a\}$ the group generators), then it must be that $R[^\theta A]$ is minimized at $\theta(x) = 0$. This requires, first of all, stationarity with

---

[1]BRST quantization is described in most modern textbooks on quantum field theory; e.g. [5].

respect to infinitesimal variations of $\theta(x)$, i.e. that

$$\left(\frac{\delta R[^\theta A]}{\delta \theta^a(x)}\right)_{|\theta=0} = \frac{1}{g}\partial_\mu A_\mu^a(x) \tag{10.4}$$

must vanish, which is the usual Landau gauge condition. Secondly, for a local minimum the Hessian

$$\left(\frac{\delta^2 R[^\theta A]}{\delta \theta^a(x)\delta \theta^b(y)}\right)_{|\theta=0} = \frac{1}{g^2}M_{xy}^{ab} \tag{10.5}$$

must have only positive semidefinite eigenvalues, where

$$M_{xy}^{ab} = -\partial_\mu D_\mu^{ab}\delta^4(x-y) \tag{10.6}$$

is the F-P operator, with $D_\mu^{ab} = \delta^{ab}\partial_\mu - gf^{acb}A^c$ the covariant derivative. From this it is clear that the Landau gauge Gribov region consists of all local minima of $R[A]$ for each gauge orbit.

The lattice version of Landau gauge, restricted to the Gribov region, is the gauge which minimizes

$$R = -\sum_x \sum_\mu \text{ReTr}[U_\mu(x)] . \tag{10.7}$$

Denoting gauge-transformed links by

$$^\theta U_\mu(x) = e^{i\theta^a(x)L_a}U_\mu(x)e^{-i\theta^a(x+\hat{\mu})L_a} , \tag{10.8}$$

the stationarity condition for $R[^\theta U]$ is

$$\left(\frac{\partial R[^\theta U]}{\partial \theta^a(x)}\right)_{|\theta=0} = i\sum_\mu \text{Tr}[L_a(U_\mu(x) + U_\mu^\dagger(x-\hat{\mu})] = 0 , \tag{10.9}$$

which becomes the usual Landau gauge condition $\partial_\mu A_\mu^a(x) = 0$ in the continuum limit. The lattice F-P operator M_{xy}^{ab} is again obtained from the Hessian matrix and, as in the continuum, the condition that the field configuration is a local *minimum* of R is that the eigenvalues of M_{xy}^{ab} are all positive (i.e. all directions on the gauge orbit away from the stationary point increase the value of R).

In Coulomb gauge, the sum over the spacetime index $\mu = 0, 1, 2, 3$ in the above expressions is replaced by a sum over a space index $k = 1, 2, 3$, and the 4-dimensional delta function in (10.6) is replaced by a 3-dimensional delta function.

In lattice simulations one generates configurations in the usual way, and observables are computed after fixing these configurations to the required gauge. But the techniques which are used to gauge fix on the lattice only obtain local minima of

$R[U]$, rather than arbitrary stationary points, and this guarantees that the eigenvalues of the F-P operator M_{xy}^{ab} are almost all positive. The exceptions are the trivial zero modes, whose origin is the remnant global symmetry $U_\mu(x) \rightarrow GU_\mu(x)G^\dagger$, where $G \in SU(N)$ is any position-independent group element. The remnant symmetry implies that at any stationary point of R there must be flat directions along the gauge orbit, corresponding to the generators of the global transformations, and these flat directions mean that there must be zero modes of the F-P operator. These are the trivial eigenstates of the F-P operator M, which are easily seen, by inspection of (10.6), to be

$$\phi_n^a(x) = \frac{1}{\sqrt{V}}\delta_{na} \quad , \quad n = 1, 2, \ldots, N^2 - 1 . \tag{10.10}$$

The statement that the F-P determinant is positive in the Gribov region really refers to the determinant of the operator in the subspace orthogonal to these trivial zero modes.

The problem with defining the path integral, in Landau or Coulomb gauge, as a sum over all copies in the Gribov region is that it cannot be guaranteed that all gauge orbits are weighted equally; it may be that different gauge orbits intersect the Gribov region different numbers of times. Ideally, one would like to restrict the region of functional integration to a subset of the Gribov region, which each gauge trajectory intersects only once. An example is the *Fundamental Modular Region* suggested by Zwanziger [3], which consists, among all local minima of R, of configurations which give the absolute minimum of R along the gauge orbit. These configurations, presumably, are unique. Unfortunately there is no known procedure for locating such absolute minima in practice, nor is there any reason to think that, if we ordered the local minima in order of their value of R, the gauge copy with the lowest minimum in the set is more physically relevant than, say, the copy with the ten-thousandth minimum in the sequence. Lattice simulations in Coulomb and Landau gauges generally ignore this issue, in the hope that the choice of gauge copy will not make a serious difference to the final results. Some simulations, however, apply a technique known as simulating annealing [6], which supplies a local minimum of R that is almost always lower than that of a gauge copy chosen at random. But it is not entirely clear, on physical grounds, why a lower minimum should be preferred.

The Gribov region is bounded in its entirety, and the Fundamental Modular region is bounded in part, by a hypersurface known as the first Gribov horizon, as illustrated in Fig. 10.1. Consider the eigenvalue equation for the F-P matrix

$$M_{xy}^{ab}\phi_y^{b(n)} = \lambda_n\phi_x^{a(n)} , \tag{10.11}$$

where summation (lattice) or integration (continuum) over the repeated y-coordinate is implied. The Gribov region, by definition, is the region in which all the λ_n are positive, apart from (in $SU(N)$) the first $N^2 - 1$ zero eigenvalues of the trivial zero modes. The *first* Gribov horizon is the boundary of the Gribov region, where the lowest non-trivial eigenmode also has a zero eigenvalue, while all other eigenvalues

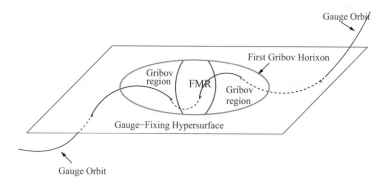

Fig. 10.1 The rectangular region indicates the subspace, in the space of all gauge configurations satisfying the (Coulomb or Landau) gauge condition. A typical gauge orbit intersects the subspace many times; the intersections (marked "x") are known as "Gribov copies." Within the Gribov region the non-trivial eigenvalues of the F-P operator are all positive; this region is bounded by the first Gribov horizon, where one non-trivial eigenvalue is zero. One can also define subregions within the Gribov region, such as the Fundamental Modular region (FMR), which are intersected only once by each gauge orbit

are positive (here I will use the terms "Gribov horizon" and "first Gribov horizon" interchangeably).

The speculation that all this has something to do with confinement is based on the fact that the Coulomb potential, as we will see below, involves the inverse of the F-P operator M. Zwanziger [3] has argued that most of the volume of the Gribov region is concentrated near the boundary, i.e. at the first Gribov horizon, in much the same way that the volume of a sphere of radius R, in d dimensions, is concentrated at $r \approx R$ for large d. Since the first non-trivial eigenvalue of M vanishes at the first Gribov horizon, it is possible that the near-zero eigenvalues of M, for configurations very near the Gribov horizon, will considerably enhance the Coulomb potential, and possibly strengthen it to a confining form. In the next sections we will see evidence that this speculation is probably correct.

10.2 Coulomb Potential on the Lattice

Coulomb gauge, like axial and temporal gauge (but *unlike* the covariant gauges) is a "physical" gauge, in the sense that it is possible to formulate a ghost-free Hamiltonian operator, and avoid the introduction of propagating ghost fields. The classical Coulomb-gauge Hamiltonian of Yang-Mills theory is $H = H_{glue} + H_{coul}$, where [7]

$$H_{glue} = \tfrac{1}{2} \int d^3x \, (\mathbf{E}_T^2 + \mathbf{B}^2)$$

$$H_{coul} = \tfrac{1}{2} \int d^3x d^3y \, \rho^a(x) K^{ab}(x, y) \rho^b(y)$$

$$K^{ab}(x, y; A) = \left[(M^{-1})^{ac}_{\mathbf{xz}}(-\nabla^2)_{\mathbf{z}}(M^{-1})^{cb}_{\mathbf{zy}} \right]$$

$$\rho^a = \rho^a_m - g f^{abc} A^b_k E^c_{Tk} ,\qquad (10.12)$$

and where E_T denotes the transverse E-field $\nabla \cdot \mathbf{E}^a_T = 0$.[2] The color-charge density due to matter fields is given by $\rho_m(x)$. If this charge density is due to static quark-antiquark color charge sources separated by a distance R, then the corresponding Coulomb interaction + self-energies are given by

$$\mathscr{E}_{Coul}(R) = \tfrac{1}{2} \int d^3x d^3y \left\langle \rho^a_m(x) K^{ab}(x, y) \rho^b_m(y) \right\rangle . \qquad (10.13)$$

In the path-integral formulation it is useful to introduce Faddeev-Popov ghost fields, but in Coulomb gauge these ghost fields do not propagate in time. Their propagator at equal times is given by $\langle (M^{-1})^{ab}_{\mathbf{xy}} \rangle$.

A glance at the operator $K(x, y; A)$ in (10.12) shows that it involves two factors of the inverse F-P operator M^{-1}. If the functional integration is dominated by configurations at or near the first Gribov horizon, then the eigenvalue spectrum of M begins at or near zero. This leads to the suggestion, long advocated by Gribov [1] and Zwanziger [3], that $\langle \rho_m K \rho_m \rangle$ is so greatly enhanced in the infrared by the near-zero eigenvalues of M that a linear potential

$$\langle K(x, y; A) \rangle \sim \sigma |\mathbf{x} - \mathbf{y}| \qquad (10.14)$$

is obtained. Since $\langle K(x, y; A) \rangle$ is essentially the instantaneous piece of the correlator $\langle A_0(x) A_0(y) \rangle$, confinement could then be understood as arising from one-gluon exchange with a (highly non-perturbative) dressed gluon propagator.

We would like to study the validity of this Gribov-Zwanziger scenario numerically, and the first step is to figure out how to compute the Coulomb energy on the lattice. Let us begin with a physical state containing a heavy quark-antiquark pair, separated by distance R:

$$|\Psi_{q\bar{q}}\rangle = \bar{q}(0) q(R) |\Psi_0\rangle , \qquad (10.15)$$

where Ψ_0 is the vacuum state of Yang-Mills theory in Coulomb gauge. Note that this is not necessarily the lowest energy state containing the two static sources, but it *is* the state in which the $\bar{q}(0)q(R)$ interaction energy is entirely Coulombic, i.e.

$$\mathscr{E}_{Coul}(R) = \langle \Psi_{q\bar{q}} | H | \Psi_{q\bar{q}} \rangle - \langle \Psi_0 | H | \Psi_0 \rangle$$

$$= V_{coul}(R) + E_{se} , \qquad (10.16)$$

---

[2]At the quantum level, the Hamiltonian contains factors of $(\det M)^{1/2}$ and $(\det M)^{-1/2}$, which cancel out in the classical limit where the ordering of operators is irrelevant [8]. These factors also cancel out in the expression for the Coulomb energy due to static color sources.

where the Coulomb potential $V_{coul}(R)$ comes from the non-local $\rho_m K \rho_m$ term in the Hamiltonian, and E_{se} is some R-independent self-energy term. We would like to know, first of all, if $V_{coul}(R)$ is confining. If it is confining, then is it asymptotically linear? If linear, does the Coulomb string tension σ_{coul} equal the string tension σ of the static quark potential? Finally, is there any connection to the center vortex mechanism? What happens to the Coulomb potential if vortices are removed?

In order to compute the Coulomb potential numerically, we begin with

$$G(R, T) = \langle \Psi_{q\bar{q}} | e^{-(H-E_0)T} | \Psi_{q\bar{q}} \rangle , \tag{10.17}$$

where E_0 is the vacuum energy. For very massive quarks in an $SU(N)$ gauge theory, by the same that reasoning that led to introducing Polyakov lines, and up to an overall constant of $O(m^{2T})$,

$$G(R, T) = \frac{1}{N} \langle \text{Tr}[L^\dagger(0, T)L(R, T)] \rangle , \tag{10.18}$$

where

$$L(\mathbf{x}, T) = P \exp \left[i \int_0^T dt \, A_0(\mathbf{x}, t) \right] \tag{10.19}$$

is a Wilson line in the time direction. Here T is some finite time lapse, but it is *not* the full extension of the lattice in the time direction. Inserting a complete set of states in (10.17)

$$G(R, T) = \sum_n \left| \langle \Psi_n | \Psi_{q\bar{q}} \rangle \right|^2 e^{-(E_n - E_0)T} . \tag{10.20}$$

Denote

$$V(R, T) = -\frac{d}{dT} \log[G(R, T)] . \tag{10.21}$$

Then it is easy to see that

$$\begin{aligned} \mathcal{E}_{coul}(R) &= \langle \Psi_{q\bar{q}} | H - E_0 | \Psi_{q\bar{q}} \rangle \\ &= V_{coul}(R) + E_{se} \\ &= V(R, 0) , \end{aligned} \tag{10.22}$$

while

$$\begin{aligned} \mathcal{E}_{min}(R) &= V(R) + E'_{se} \\ &= \lim_{T \to \infty} V(R, T) , \end{aligned} \tag{10.23}$$

where $\mathcal{E}_{min}$ is the lowest energy that can be obtained among states which contain the static $q\bar{q}$ system, and $V(R)$ is the static quark potential. In general there is no reason that the minimal energy state should be the $\Psi_{q\bar{q}}$ state defined in (10.15), which is used to extract the Coulomb potential. With lattice regularization, E_{se} and E'_{se} are negligible at large R, compared to $V(R)$. Then, since $\mathcal{E}_{min} \leq \mathcal{E}_{coul}$, it follows that

$$V(R) \leq V_{coul}(R) , \tag{10.24}$$

as first noted by Zwanziger in [9]. Therefore, if confinement exists at all, there must also be a confining Coulomb potential.

On the lattice

$$L(x, T) = U_0(x, 1)U_0(x, 2) \cdots U_0(x, T)$$

$$V(R, T) = \frac{1}{a} \log \left[\frac{G(R, T)}{G(R, T + 1)} \right] . \tag{10.25}$$

Then

$$V(R, 0) = V_{coul}(R) + \text{const.}$$

$$\lim_{T \to \infty} V(R, T) = V(R) + \text{const.} , \tag{10.26}$$

where, using the fact that $L(\mathbf{x}, 0) = \mathbb{1}$ by definition,

$$V(R, 0) = -\log[G(R, 1)] \tag{10.27}$$

in lattice units. This relation allows us to derive the lattice Coulomb potential ($= V(R, 0)$ up to a constant) from the timelike link-link correlator $G(R, 1)$ at equal times, and compare to the static potential $V(R)$ on the lattice.

We may also consider the Coulomb energy of a one-particle state $\Psi_q^a = q^a(x)\Psi_0$. By the same reasoning as above, the Coulomb energy computed on the lattice will be

$$\mathcal{E}_q \propto -\log[\langle \text{Tr}[U_0(x)]\rangle] , \tag{10.28}$$

and therefore the Coulomb energy of this one-particle state is infinite if the expectation value of a timelike link vanishes in Coulomb gauge. Recall that this was also the condition, discussed in Chap. 3, that the remnant gauge symmetry $U_0(\mathbf{x}, t) \to g(t)U_0(\mathbf{x}, t)g^\dagger(t + 1)$ is unbroken. It should be understood, however, that $\mathcal{E}_q$ is the energy of a specific state, and as such it is only an upper bound on the minimal energy of a states containing a single static quark. Therefore $\mathcal{E}_q = \infty$ is a necessary but not sufficient condition for the non-existence of finite-energy isolated charge states.

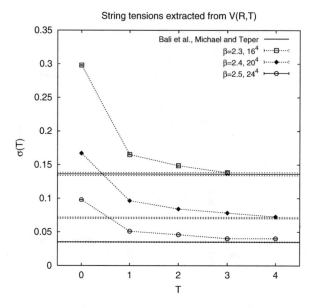

Fig. 10.2   Falloff of $\sigma(T)$ with increasing T at $\beta = 2.3, 2.4, 2.5$. Solid lines indicate the accepted values of the asymptotic string tension at each β value, with dashed lines indicating the error bars. From [10]

Returning to the lattice Coulomb potential for static $q\bar{q}$ sources, we begin with a check: Define $\sigma(T)$ from a fit of $V(R, T)$ to

$$V(R, T) = c(T) - \frac{\pi}{12R} + \sigma(T)R , \qquad (10.29)$$

where $-\pi/12R$ is the Lüscher term, and see if $\sigma(T) \to \sigma$ in the large-T limit. This seems to work out as expected, as seen in Fig. 10.2. Figure 10.3 shows the data for $V(R, 0)$, at $\beta = 2.5$, which can be identified with the instantaneous Coulomb potential up to some R-independent self-energy tern.[3] The figure also displays the result (lower line) obtained from lattice configurations with vortices removed by the de Forcrand-D'Elia procedure, which was introduced in Sect. 6.3.3.

The potential $V(R, T)$ contains a self-energy term $c(T)$ in lattice units, which diverges to infinity when $V(R, T)$ is converted to physical units. Subtracting the self-energy we convert to physical units by dividing by the lattice spacing $a(\beta)$, and multiplying by a conversion factor $1 = 0.197$ Gev-fm to express the Coulomb

---

[3] Another approach is to calculate $\langle K^{ab}(x, y; A) \rangle$ directly, via Monte Carlo simulations. This computationally more demanding procedure has been followed in [11] for the $SU(3)$ gauge group, with the result that the Coulomb potential is linearly confining, with a string tension that is greater (by a factor estimated at 1.6) than the asymptotic string tension.

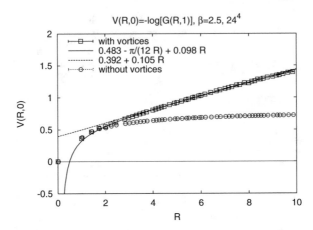

Fig. 10.3 $V(R, 0)$ at $\beta = 2.5$ in SU(2) lattice gauge theory, which is related (by (10.22)) to the Coulomb energy. The solid (dashed) line is a fit to a linear potential with (without) the Lüscher term. From [10]

interaction energy at a particular lattice coupling β in physical units, i.e. we define

$$V^{int}(R_{phys}, \beta) = \frac{0.197}{a(\beta)}(V(R, 0) - c(0)) \qquad (10.30)$$

where $R_{phys} = Ra(\beta)$ is in units of fermis, and V^{int} is in GeV. For SU(3) lattice gauge theory the lattice spacing is given by the Necco-Sommer formula [12]

$$a = r_0 \exp\left(-1.6804 - 1.7331(\beta - 6)\right.$$
$$\left. +0.7849(\beta - 6)^2 - 0.4428(\beta - 6)^3\right), \qquad (10.31)$$

with $r_0 = 0.5$ fm. The crucial test is whether the data for the Coulomb interaction energy $V^{int}(R_{phys}, \beta)$ converges as $\beta \to \infty$.

This question was investigated in [13] and [14] with results shown in Fig. 10.4. There does appear to be convergence in V^{int} as β increases. The data in physical units was also fit to the form

$$V^{int}(R_{phys}, \beta) = \sigma_c(\beta)R_{phys} - (0.197 \text{ GeV-fm})\frac{\gamma(\beta)}{R_{phys}}, \qquad (10.32)$$

and it was observed that both $\sigma_c(\beta)$ and $\gamma(\beta)$ appear to converge in the $\beta \to \infty$, $a(\beta) \to 0$ limit. It is interesting that $\gamma(\beta)$ may very well be converging to the value expected from the Lüscher term, i.e. $\gamma = \pi/12 = 0.262$, as seen in Fig. 10.5. The values of the Coulomb string tension in physical units at various β couplings are shown in Fig. 10.6. From the last data point at $\beta = 6.4$, $a = 0.051$ fm, we

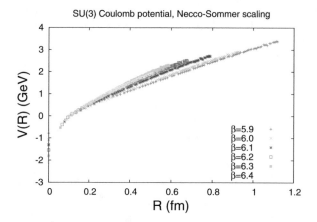

Fig. 10.4 The instantaneous Coulomb potential in physical units, for a range of lattice couplings β, with self-energies subtracted as explained in the text. From [13]

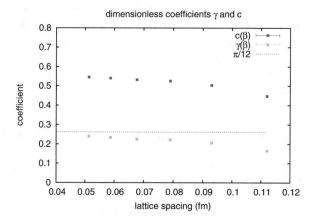

Fig. 10.5 The self-energy term $c(\beta)$, and the coefficient $\gamma(\beta)$ of the $1/R_L$ term in the instantaneous Coulomb potential (derived from fits to (10.29) and (10.32), respectively), vs. lattice spacing $a(\beta)$. The flat line at $\pi/12$ indicates the value of the coefficient of the Lüscher term. Error bars are smaller than the symbol size. From [13]

estimate the Coulomb string tension to be $\sigma_c \approx 4.03(8)$ GeV/fm, or in other units $\sigma_c = (891 \pm 9\,\text{MeV})^2$, to be compared to the accepted value of $\sigma = (440\,\text{MeV})^2$ for the asymptotic string tension. These values are differ by more than a factor of four. While this is perfectly consistent with the Zwanziger inequality (10.24), there is still a question of what mechanism reduces the Coulomb string tension, by such a large factor, to the asymptotic value. The gluon chain model, presented in Chap. 12, has tried to address this question. There have also been some recent studies of the color Coulomb electric field strength distribution around static quark-antiquark sources,

Fig. 10.6 The Coulomb
string tension σ_c in physical
units (Gev/fm) vs. lattice
spacing $a(\beta)$. Error bars are
smaller than the symbol size.
From [13]

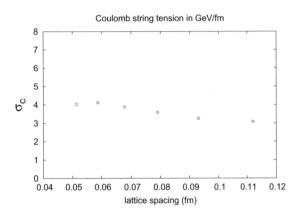

with evidence of some degree of flux collimation, but the results at this stage do not
appear to be definitive, c.f. [15, 16].

We have seen in an earlier chapter that the asymptotic string tension vanishes
upon vortex removal, and it is clear from Fig. 10.3 that the Coulomb string tension
also vanishes in vortex-removed configurations. This brings up two interesting
questions. First, what property of Yang-Mills vacuum configurations is responsible
for the confining Coulomb potential? Secondly, how is this property related to the
existence of center vortices?

10.3   F-P Eigenvalue Density, and the Coulomb Self-Energy

In a confining theory, the energy of a color-nonsinglet state is infinite. In Coulomb
gauge, such a state is, e.g.

$$\Psi^a[A; x] = q^a(x)\Psi_0[A] . \tag{10.33}$$

There is nothing intrinsically unphysical (or gauge non-invariant) about such a
state, at least in an infinite volume. Gauge invariance in a Hamiltonian formulation
amounts to satisfying the Gauss Law in one way or another. In Coulomb gauge
only physical degrees of freedom appear, and Gauss's Law is solved explicitly,
resulting in the H_{coul} term in the Hamiltonian. In $A_0 = 0$ gauge, Gauss's Law
is an operator constraint on physical states, requiring that such states are invariant
under infinitesimal gauge transformations. In QED, the state in $A_0 = 0$ gauge which
corresponds to the state (10.33) in Coulomb gauge is

$$\Psi[A; x] = \exp\left[ie \int d^3z \, A(z) \cdot \nabla \frac{1}{4\pi|x - z|}\right] q(x)\Psi_0[A] . \tag{10.34}$$

This conversion of the charged Coulomb gauge state to temporal gauge can be
extended to non-abelian theories [17].

While the state corresponding to an isolated color non-singlet particle (10.33) qualifies as a physical state in Coulomb gauge, we would nevertheless expect that the self-energy of such a state is infinite in a confining theory, even with an ultraviolet cutoff. Before studying this Coulombic self-energy in Yang-Mills theory, it is instructive to do the calculation in an abelian theory, but in a way which will easily generalize to the non-abelian case. In the abelian theory we have

$$H = \int d^3x \; (\mathbf{E}_T^2 + (\nabla \times \mathbf{A})^2) \; + \; H_{coul}$$

$$H_{coul} = \tfrac{1}{2} \int d^3x d^3y \; \rho(x) K(x, y) \rho(y)$$

$$K(x, y) = M_{xz}^{-1}(-\nabla^2)_z M_{zy}^{-1} \; , \tag{10.35}$$

and also

$$\mathcal{E} = e^2 K(x, x) \; , \tag{10.36}$$

where $\mathcal{E}$ is the Coulomb self-energy of a static charge at point x. As in the non-abelian theory, the abelian F-P operator is obtained from the stationarity of the integral over $A_\mu A_\mu$, but in this case M is independent of the gauge field

$$M_{xy} = -\nabla^2 \delta(x - y) \; . \tag{10.37}$$

The eigenstates of the F-P operator, satisfying

$$M_{xy}\phi_y^{(n)} = \lambda_n \phi_x^{(n)} \; , \tag{10.38}$$

are simply the plane wave states, with eigenvalues $\lambda_n = \mathbf{k}_n \cdot \mathbf{k}_n$ for wavenumber $\mathbf{k}_n$. These states are discrete in a finite volume, and integration over position become summation over position in a lattice regularization. We can write the Green's function in the form

$$G_{xy} = \left[M^{-1}\right]_{xy} = \sum_n \frac{\phi_x^{(n)} \phi_y^{(n)*}}{\lambda_n} \; . \tag{10.39}$$

With some simple manipulations, on a lattice of spatial volume $V = L^3$,

$$\mathcal{E} = \frac{e^2}{L^3} \left(M^{-1}(-\nabla^2)M^{-1}\right)_{xx}$$

$$= \frac{e^2}{L^3} \sum_x \sum_{y_1 y_2} G_{xy_1}(-\nabla^2)_{y_1 y_2} G_{y_2 x}$$

$$= \frac{e^2}{L^3} \sum_x \sum_{y_1 y_2} \sum_m \sum_n \frac{\phi_x^{(m)} \phi_{y_1}^{(m)*}}{\lambda_m} (-\nabla^2)_{y_1 y_2} \frac{\phi_{y_2}^{(n)} \phi_x^{(n)*}}{\lambda_n}$$

$$= \frac{e^2}{L^3} \sum_{y_1 y_2} \sum_n \frac{\phi_{y_1}^{(n)*} (-\nabla^2)_{y_1 y_2} \phi_{y_2}^{(n)}}{\lambda_n^2} \,, \tag{10.40}$$

we can express the Coulomb self-energy in the form

$$\mathcal{E} = \frac{e^2}{L^3} \sum_n \frac{F(\lambda_n)}{\lambda_n^2} \,, \tag{10.41}$$

where

$$F(\lambda_n) = \left(\phi^{(n)} \middle| (-\nabla^2) \middle| \phi^{(n)} \right) . \tag{10.42}$$

Let $\rho(\lambda)$ denote the density of eigenvalues, scaled so that

$$\int d\lambda \, \rho(\lambda) = 1 . \tag{10.43}$$

Then at large volumes the sum over eigenstates can be approximated by an integral

$$\mathcal{E} = e^2 \int d\lambda \, \frac{\rho(\lambda) F(\lambda)}{\lambda^2} . \tag{10.44}$$

In QED it is easy to show (see Eq. (6.59) with $D = 3$) that

$$\rho(\lambda) = \frac{\sqrt{\lambda}}{4\pi^2} \,, \quad F(\lambda) = \lambda . \tag{10.45}$$

The minimum eigenvalue in a volume of extension L is $\lambda_{min} \sim 1/L^2$, and the maximum eigenvalue $\lambda_{max} \sim 1$. Putting it all together, we find (in lattice units)

$$\mathcal{E} = e^2 \int d\lambda \, \frac{\rho(\lambda) F(\lambda)}{\lambda^2}$$

$$\sim e^2 \left(1 - \frac{1}{L} \right) . \tag{10.46}$$

In physical units (divide the lattice expression by lattice spacing a), this energy has the familiar UV divergence as $a \to 0$. The second term is a finite volume effect, which goes to zero in the infinite volume limit. However, finiteness at $L \to \infty$ at fixed a clearly depends on the small λ behavior of $\rho(\lambda)F(\lambda)$. If we had instead

$$\lim_{\lambda \to 0} \frac{\rho(\lambda)F(\lambda)}{\lambda} > 0 \,, \tag{10.47}$$

then the Coulomb energy would be divergent in the large volume limit.

Typical configurations in the Gribov region are expected to approach the Gribov horizon in the infinite-volume limit. This is true even at the perturbative level, where $\lambda_{min} \sim 1/L^2$. But what counts for confinement is the density of eigenvalues $\rho(\lambda)$ near $\lambda = 0$, and the lack of smoothness of these near-zero eigenvalues, as measured by $F(\lambda)$. This is what determines whether the Coulomb confinement criterion (10.47) is satisfied.

In Yang-Mills theory the calculation of the self-energy of a static charge goes in much the same way, except that the F-P operator (10.6) depends on the gauge field, and for an arbitrary gauge group, with static charge in color representation r,

$$\mathscr{E}_r = \frac{g^2 C_r}{d_A}\mathscr{E} \,, \tag{10.48}$$

where C_r is the quadratic Casimir of representation r, d_A is the dimension of the adjoint representation, and

$$\mathscr{E} = \int d\lambda \, \left\langle \frac{\rho(\lambda)F(\lambda)}{\lambda^2} \right\rangle \,,$$

$$F(\lambda_n) = \left(\phi^{(n)} \middle| (-\nabla^2) \middle| \phi^{(n)} \right) \,. \tag{10.49}$$

In order to test if $\mathscr{E}$ is infrared divergent, we can calculate $\rho(\lambda)$, $F(\lambda)$, numerically, on finite-size lattices, and extrapolate to infinite volume. This was done in [18], at a coupling $\beta = 2.1$ corresponding to a rather large lattice spacing, which allows us to probe fairly large volumes in physical units. The procedure is to generate field configurations by the lattice Monte Carlo method, and fix each of them to Coulomb gauge (i.e. to a gauge copy in the Gribov region). Then, for each configuration, the first 200 eigenstates of the lattice F-P operator on each lattice time-slice are computed numerically, via the Arnoldi algorithm [19]. From these we calculate the average, over all gauge-fixed configurations, of $\rho(\lambda)$, $F(\lambda)$. The results, for lattice sizes 8^4 to 20^4 are shown in Fig. 10.7 for $\rho(\lambda)$, and Fig. 10.8 for $F(\lambda)$. From the scaling of this data with lattice size at small λ (cf. [18] for details), it is found that

$$\rho(\lambda) \sim \lambda^{0.25} \,, \quad F(\lambda) \sim \lambda^{0.4} \,, \tag{10.50}$$

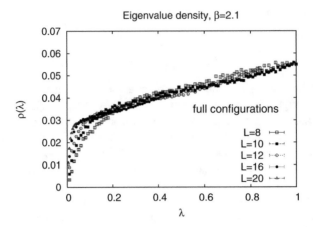

Fig. 10.7 The F-P eigenvalue density at $\beta = 2.1$, on 8^4–20^4 lattice volumes. From [18]

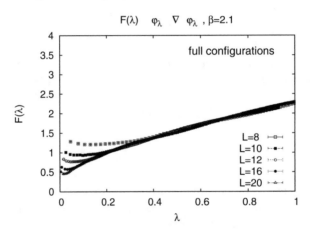

Fig. 10.8 $F(\lambda)$, the diagonal matrix element of $(-\nabla^2)$ in F-P eigenstates, plotted vs. F-P eigenvalue. From [18]

and therefore

$$\int d\lambda \, \frac{\rho(\lambda)F(\lambda)}{\lambda^2} \to \infty \tag{10.51}$$

is divergent in the infinite volume limit, due to the degree of divergence of the integrand at the lower ($\lambda = 0$) end of the integration region. This means that the Coulomb energy of an isolated color charge, in the state represented by Ψ^a in (10.33), is infinite, even when the usual ultraviolet divergence is regulated on the lattice.

10.3.1 The Role of Center Vortices

Next we can ask whether center vortices are related, in some way, to the divergent Coulomb energy of an isolated charge. The simplest thing to do is to repeat the above calculation for center-projected configurations ("vortex-only"), and for configurations with the vortices removed via the de Forcrand-D'Elia prescription [20]. Figures 10.9 and 10.10 show the results for $\rho(\lambda)$ and $F(\lambda)$ in the center-projected lattices. From a lattice size scaling analysis it is found that, at small λ,

$$\rho(\lambda) \sim \lambda^{0\pm0.05} \quad , \quad F(\lambda) \approx 1 \, , \tag{10.52}$$

and again the expression (10.51) is divergent.

Figures 10.11 and 10.12 (left) show the corresponding results for $\rho(\lambda)$ and $F(\lambda)$ in vortex-removed configurations. This time there is a drastic change, as compared to the unmodified, and center-projected lattices; the density of states shows a series of peaks, and $F(\lambda)$ has a "band" structure. The peaks and bands correspond, in the sense that the eigenvalues belonging to the same peak of ρ also belong to the same band of F.

It turns out that the structure of peaks in $\rho(\lambda)$ is essentially perturbative in origin. On a finite lattice, the density of states of the ordinary Laplacian $-\nabla^2$ (the $g = 0$ approximation to the covariant Laplacian) is a series of delta functions, each of which is singular at one of the eigenvalues of the Laplacian. The degeneracy of each eigenvalue is computable, and one finds that the numbers of eigenvalues in the peaks of $\rho(\lambda)$ precisely match the degeneracies of the low-lying eigenvalues of $-\nabla^2$. The effect of the vortex-removed lattice fields is to broaden the delta functions into peaks of finite width. The impression that the vortex-removed configurations are only a perturbation of the free theory is strengthened by the fact that the lowest-lying eigenvalues go to zero like $1/L^2$, where the L is the lattice extension, just as

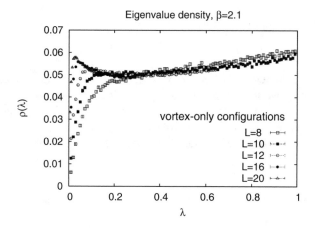

Fig. 10.9 F-P eigenvalue density in vortex-only configurations. From [18]

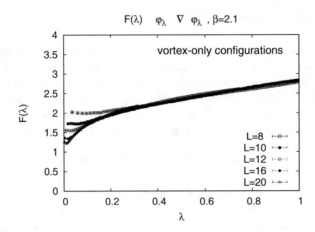

Fig. 10.10 $F(\lambda)$, the diagonal matrix element of $(-\nabla^2)$ in F-P eigenstates, for vortex-only configurations. From [18]

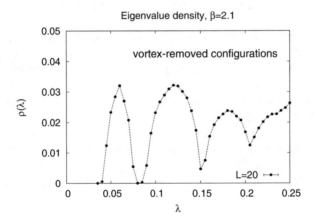

Fig. 10.11 F-P eigenvalue densities for vortex-removed configurations, on a 20^4 lattice volume. From [18]

in a free theory. Moreover, if we plot our data for $F(\lambda)$ vs. λ at all volumes, as in Fig. 10.12 (right), we see that $F(\lambda) \propto \lambda$, again as in a free theory.

From all of this data, it appears that the enhancement of $\rho(\lambda)$ and $F(\lambda)$ near $\lambda = 0$, which is responsible for the infrared divergent Coulomb self-energy of an isolated charge, is correlated in some way with the presence of a center vortex ensemble.

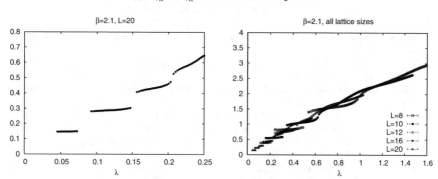

Fig. 10.12 $F(\lambda)$ vs. λ in the vortex-removed configurations. We display results for the 20^4 volume alone (left), and a variety of lattice volumes (right). From [18]

10.4 Critique

Although all available numerical evidence favors a confining Coulomb force, and indeed this is a necessary condition for the existence of a confining static quark potential, there are also good reasons to doubt that the confining force between static color charges is simply the Coulomb force in Coulomb gauge. In the first place, a necessary condition is not a sufficient condition, and there is also numerical evidence that the Coulomb force is confining at high temperatures, past the deconfinement transition [21]. We have also seen that even at zero temperature, the Coulomb string tension is several times larger than the asymptotic string tension. Nevertheless, it is possible that the confining Coulomb potential is an important element in understanding the formation of confining flux tubes, because this formation might be understood as a response which brings the Coulombic string tension down to the asymptotic value. The gluon chain model, to be discussed in Chap. 12, is one proposal along these lines.

References

1. V. Gribov, Quantization of non-Abelian gauge theories. Nucl. Phys. B **139**, 1 (1978)
2. D. Zwanziger, Vanishing of the zero-momentum lattice gluon propagator and color confinement. Nucl. Phys. B **364**, 127 (1991)
3. D. Zwanziger, Renormalization in Coulomb gauge and an order parameter for confinement in QCD. Nucl. Phys. B **518**, 237 (1998)
4. H. Neuberger, Non-perturbative BRS invariance and the Gribov problem. Phys. Lett. B **183**, 337 (1987)
5. M. Srednicki, *Quantum Field Theory* (Cambridge University Press, New York, NY, 2007); M. Peskin, D. Schroeder, *An Introduction to Quantum Field Theory* (Perseus Books, Cambridge, MA, 1995)

6. W.H. Press, S.A. Teukolsky, W.T. Vetterling, B.P. Flannery, *Numerical Recipes: The Art of Scientific Computing* (Cambridge University Press, New York, NY, 2007)
7. E.S. Abers, B.W. Lee, Gauge theories. Phys. Rept. **9**, 1 (1973)
8. N.H. Christ, T.D. Lee, Operator ordering and Feynman rules in gauge theories. Phys. Rev. D **22**, 939 (1980)
9. D. Zwanziger, No confinement without Coulomb confinement. Phys. Rev. Lett. **90**, 102001 (2003). arXiv: hep-th/0209105
10. J. Greensite, Š. Olejník, Coulomb energy, vortices, and confinement. Phys. Rev. D **67**, 094503 (2003). arXiv: hep-lat/0302018
11. A. Voigt, E.M. Ilgenfritz, M. Muller-Preussker, A. Sternbeck, The effective Coulomb potential in SU(3) lattice Yang-Mills theory. Phys. Rev. D **78**, 014501 (2008). arXiv:0803.2307 [hep-lat]
12. S. Necco, R. Sommer, The N(f) = 0 heavy quark potential from short to intermediate distances. Nucl. Phys. B **622**, 328 (2002). hep-lat/0108008
13. J. Greensite, A.P. Szczepaniak, Coulomb string tension, asymptotic string tension, and the gluon chain. Phys. Rev. D **91**(3), 034503 (2015). arXiv:1410.3525 [hep-lat]
14. Y. Nakagawa, A. Nakamura, T. Saito, H. Toki, D. Zwanziger, Properties of color-Coulomb string tension. Phys. Rev. D **73**, 094504 (2006). hep-lat/0603010
15. K. Chung, J. Greensite, Coulomb flux tube on the lattice. Phys. Rev. D **96**(3), 034512 (2017). https://doi.org/10.1103/PhysRevD.96.034512. arXiv:1704.08995 [hep-lat]
16. S.M. Dawid, A.P. Szczepaniak, The Coulomb flux tube revisited. Phys. Rev. D **100**, 074508 (2019). https://doi.org/10.1103/PhysRevD.100.074508
17. M. Lavelle, D. McMullan, Hadrons without strings. Phys. Lett. B **471**, 65 (1999). arXiv: hep-ph/9910398
18. J. Greensite, S. Olejnik, D. Zwanziger, Center vortices and the Gribov horizon. JHEP **0505**, 070 (2005). arXiv:hep-lat/0407032
19. R.B. Lehoucq, D.C. Sorensen, C. Yang, *Arpack User's Guide: Solution of Large-Scale Eigenvalue Problems with Implicitly Restarted Arnoldi Methods* (SIAM, 1998)
20. P. de Forcrand, M. D'Elia, On the relevance of center vortices to QCD. Phys. Rev. Lett. **82**, 4582 (1999). arXiv:hep-lat/9901020
21. J. Greensite, S. Olejnik, D. Zwanziger, Coulomb energy, remnant symmetry, and the phases of non-Abelian gauge theories. Phys. Rev. D **69**, 074506 (2004). arXiv:hep-lat/0401003

Ghosts, Gluons, and Dyson-Schwinger Equations

11

Shortly after the recognition, in the 1970s, that the static quark potential ought to be linear with quark separation, there was an effort to show that the momentum space gluon propagator, in covariant gauges, goes as $1/k^4$ as $k \to 0$. The reason this was considered desirable is that if we naively evaluate the static quark potential from one-gluon exchange, then it is not hard to show that a $1/k^4$ limit at low momenta leads to a linear potential. Attempts to derive this $1/k^4$ behavior were based on trying to solve a truncated set of Dyson-Schwinger equations (DSEs), but because of various ambiguities, and the doubtful validity of the truncation, the effort was abandoned after a few years.

In the late 1990s a series of articles by Alkofer and co-workers [1] revived interest in the Dyson-Schwinger approach, and there has been much work on the topic since then (cf. the review article by Fischer [2]). In the new approach the gluon propagator was not found to go as $1/k^4$; in fact the prediction was that the gluon propagator actually vanishes in the $k \to 0$ limit, and it is instead the ghost propagator which should have an enhanced singularity (i.e. more singular than $1/k^2$) in the infrared. This solution of the Dyson-Schwinger equations is referred to as the *scaling solution*. The enhanced infrared singularity of the ghost propagator, found in the scaling solution, is necessary for the Kugo-Ojima confinement criterion (3.12) to be satisfied [3]; it is also the "horizon condition" advocated (together with the vanishing of the gluon propagator at $k = 0$) by Zwanziger [4]. In recent years there has been a concerted effort to check these predictions by calculating ghost and gluon propagators numerically, on the lattice. Surprisingly, the lattice simulations appear to support the scaling solution in Landau gauge in $D = 2$ dimensions [5], but not in $D = 3$ and 4 dimensions [6, 7]. Further investigation has shown that there exists another consistent solution of the Dyson-Schwinger equations, known as *decoupling solution* [8], and this solution does appear to be in harmony with the $D = 3$ and 4 dimensional lattice results.

© Springer Nature Switzerland AG 2020
J. Greensite, *An Introduction to the Confinement Problem*, Lecture Notes in Physics 972, https://doi.org/10.1007/978-3-030-51563-8_11

The DSE effort is ongoing, and not everyone is in agreement that the current numerical results are decisive in favoring the decoupling solution. At the very least, we are seeing a very intriguing interplay between analytical arguments and numerical simulations, and much has been learned, in the last few years, about the infrared behavior of ghost and gluon propagators on the lattice. In this chapter I will briefly review the main developments in this area.

11.1 Dyson-Schwinger Equations and the Scaling Solution

Dyson-Schwinger equations are derived in many standard textbooks on quantum field theory (see, e.g., [9]); they follow from the fact that the functional integral of a total derivative is zero. Consider any field theory involving a set of fields $\{\phi_k\}$. Then

$$0 = \frac{1}{Z} \int \left\{ \prod_k D\phi_k \right\} \frac{\delta}{\delta \phi_i(x)} \exp\left[-S + \int d^D x \sum_i j_i(x)\phi_i(x) \right]$$

$$= \left\langle -\frac{\delta S}{\delta \phi_i(x)} + j_i(x) \right\rangle . \tag{11.1}$$

Further differentiation with respect to the sources $\{j_k\}$ yields the Dyson-Schwinger equations.[1] As an example, the Dyson-Schwinger equation for the ghost propagator is derived from setting $\phi_i(x)$ equal to the antighost field $\bar{c}(x)$ and then differentiating with respect to the corresponding current source at point y, yields the Dyson-Schwinger equation for the ghost propagator

$$\left\langle \frac{\delta S_{gf}}{\delta \bar{c}^a(x)} \bar{c}^b(y) \right\rangle = \delta^{ab}\delta^D(x-y) , \tag{11.2}$$

where S_{gf} is the gauge-fixing part of the action of a non-abelian gauge theory quantized in a covariant gauge. With the help of some manipulations that can be found in Ref. [10], the above equation can be brought into a form which is easiest to indicate diagramatically, in Fig. 11.1. The left hand side of the illustrated equation is the inverse of the full (or "dressed") ghost propagator, the line on the right hand side is the inverse of the bare ghost propagator, and the rightmost loop diagram contains dressed ghost and gluon propagators connected by one bare and one dressed ghost-gluon vertex. This is one of the simplest of the Dyson-Schwinger equations, the other being the equation for the ghost-gluon vertex shown in Fig. 11.2. The important insight of Ref. [1] is that if the momentum on either of the external ghost lines of the loop diagram in Fig. 11.2 goes to zero, and the loop integral is

---

[1]In a gauge theory quantized in Landau or Coulomb gauge, where the functional integration is restricted to the Gribov region, the integration does not result in a boundary term, because the Fadeev-Popov determinant vanishes on the boundary (i.e. on the Gribov horizon).

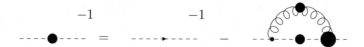

Fig. 11.1 Diagrammatic representation of the Dyson-Schwinger equation for the ghost propagator. Lines with (without) filled circles represent dressed (bare) propagators, the dashed lines are ghost propagators (or inverse propagators), and the wavy line is a dressed gluon propagator. The loop diagram contains one bare and one dressed ghost-gluon vertex. From Fischer [2]

Fig. 11.2 Diagrammatic representation of the Dyson-Schwinger equation for the ghost-gluon vertex, with the same conventions as in Fig. 11.1. From Fischer [2]

finite, then the whole diagram vanishes. This means that in the infrared limit the dressed ghost-gluon vertex should resemble the bare vertex.

The simplicity of the ghost-gluon vertex was then used in the Dyson-Schwinger equation for the ghost propagator, to derive an interesting relationship between the infrared behavior of the ghost and gluon propagators in the infrared limit. Let us write the full ghost and gluon propagators in the form

$$G^{ab}(q) = -\delta^{ab}\frac{J(q^2)}{q^2} ,$$

$$D^{ab}_{\mu\nu}(q) = \delta^{ab}\left(\delta_{\mu\nu} - \frac{q_\mu q_\nu}{q^2}\right)\frac{Z(q^2)}{q^2} , \qquad (11.3)$$

and suppose that, as $q \to 0$

$$J(q^2) \propto (q^2)^{-\kappa_{ghost}} , \qquad Z(q^2) \propto (q^2)^{\kappa_{gluon}} . \qquad (11.4)$$

If the external momenta are very small in the Dyson-Schwinger equation for the ghost propagator, then the loop integral is dominated by momenta of the same magnitude as the external momenta, and one can replace the dressed propagators by their assumed power law behavior in the infrared. Matching powers of external momenta leads to a coupling of the ghost and gluon exponents

$$\kappa_{gluon} = 2\kappa_{ghost} . \qquad (11.5)$$

This kind of analysis has been extended to other n-point functions. The result is as follows: Let $\Gamma^{(n,m)}(q^2)$ represent the dressing function for a one-particle irreducible Green's function with $2n$ external ghost lines and m external gluon lines, and all external momenta set equal to q. Then, as $q \to 0$,

$$\Gamma^{(n,m)}(q) \propto (q^2)^{(n-m)\kappa_{ghost}} . \tag{11.6}$$

The exponent κ_{ghost} has been determined from a further analysis of the gluon Dyson-Schwinger equation to have the value

$$\kappa_{ghost} = \frac{93 - \sqrt{1201}}{98} \approx 0.595 . \tag{11.7}$$

For a derivation of these relations, cf. [11]. With this value of κ_{ghost}, the scaling solution is making a clear prediction that the ghost propagator is more singular, in the infrared, than a simple $1/q^2$ pole, while the gluon propagator should actually vanish as $q \to 0$. This prediction for the ghost propagator is in perfect harmony with the Kugo-Ojima confinement criterion, which also requires an infrared singular behavior for the ghost dressing function. Given the infrared behavior of propagators and vertices, one can calculate the various running couplings associated with three gluon, four gluon, and ghost-gluon vertices, and these all run to finite non-zero values in the infrared.

One might wonder how a gluon propagator which vanishes at zero momentum, and running couplings which are finite in the infrared, will ever produce a confining static quark potential. Alkofer, Fischer, Llanes-Estrada and Schwenzer [12] have addressed this question, and they argue that the quark-gluon vertex, unlike the ghost-gluon vertex, is singular in the infrared, and that this singularity overwhelms the suppression factor of the gluon propagator to produce a confining potential. Specifically, the prediction is that at low momentum

$$\Gamma^{qg}(q^2) \sim (p^2)^{-0.5-\kappa_{ghost}} \quad , \quad Z_f(p^2) \sim \text{const.} \quad , \quad Z(p^2) \sim (p^2)^{2\kappa_{ghost}} , \tag{11.8}$$

where Γ^{qg} is the quark-gluon vertex, and Z_f is the dressing function for the quark propagator. This leads to a running coupling

$$\alpha^{qg}(q^2) \sim [\Gamma^{qg}(q^2)]^2 [Z_f(q^2)]^2 Z(q^2) \sim \frac{1}{q^2} . \tag{11.9}$$

Figure 11.3 shows the skeleton expansion of the four-quark Green's function. Already the first term has a $1/q^4$ behavior, leading to a linear potential. Note the contrast to earlier treatments, which tried to derive from the Schwinger-Dyson equations a gluon propagator behaving as $1/q^4$. In the scaling solution, it is the combination of the quark-gluon vertices with the (non-singular) gluon propagator that produces a $1/q^4$ behavior, which implies a linear potential.

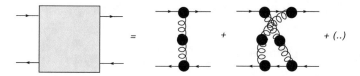

Fig. 11.3 The four quark one-particle irreducible Green's function. From Alkofer et al. [12]

One objection that could be made at this point is that the same kind of analysis seems inevitably to lead to long range interactions, i.e. van der Waals forces, between color singlet bound states. If the combination of quark-gluon vertices and gluon propagators leads to a long range interaction between color non-singlets, it would seem that the same combination would generate long-range forces also between color singlets. Such a long-range force would contradict the existence of a mass gap, which requires that there are only short-range forces between color singlet objects. It has not yet been shown that the scaling solution is consistent with this condition. A closely related issue is that it is not at all obvious, in this analysis, that a color electric flux tube forms between heavy, widely separated quarks. But however serious these problems may be, the scaling solution must first contend with a more direct challenge, coming from the lattice data.

11.2   Numerical Results for Ghost and Gluon Propagators

In view of the strong and remarkable predictions derived from the Dyson-Schwinger equations, there has naturally been an effort to check them via lattice Monte Carlo simulation. The simulations have been carried out independently, on very large lattices, by Cucchieri and Mendes [6], and by Bogolubsky et al. [7]. Both groups report similar results, and these appear to contradict the predictions of the scaling solution.

The gluon propagator on the lattice is defined in the usual way:

$$D^{ab}_{\mu\nu}(q) = \langle \tilde{A}^a_\mu(k) \tilde{A}^b_\nu(-k) \rangle$$

$$= \left(\delta_{\mu\nu} - \frac{q_\mu q_\nu}{q^2} \right) \delta^{ab} D(q^2) , \qquad (11.10)$$

where $q_\mu = (2/a) \sin(\pi k_\mu/L)$ with $k_\mu \in [-\frac{1}{2}L, \frac{1}{2}L]$. The variable $\tilde{A}^a_\mu(k)$ is the Fourier transform, in a finite lattice volume, of the gauge field defined from link variables

$$A^a_\mu(x) = \frac{1}{2iag}(U_\mu(x) - U^\dagger_\mu(x)) . \qquad (11.11)$$

The lattice ghost propagator, in momentum space, is given by

$$G^{ab}(q) = a^2 \left\langle \sum_{x,y} e^{-2\pi ik\cdot(x-y)/L} \left(M^{-1}\right)^{ab}_{xy} \right\rangle$$

$$= \delta^{ab} G(q) \,, \tag{11.12}$$

where M^{ab}_{xy} is the lattice F-P operator in Landau gauge. The calculation of $D(q)$ and $G(q)$ in lattice Landau gauge has been carried out for both the SU(2) gauge group, in $D = 3$ and $D = 4$ dimensions, and also for SU(3) gauge theory, in $D = 4$ dimensions. The results in all cases are quite similar; here we will just display the results of Bogolubsky et al. [7] for the SU(3) gauge group in $D = 4$ dimensions, at Wilson lattice coupling $\beta = 5.7$, with lattice volumes up to 96^4. Using the appropriate lattice spacing $a = 0.17$ fm at $\beta = 5.7$, all quantities can be expressed in physical units.

The result for the gluon propagator is shown in Fig. 11.4. There is no sign that the gluon propagator vanishes at zero momentum; rather it appears to reach a plateau. The ghost dressing function $J(q^2) = q^2 G(q^2)$ is displayed in Fig. 11.5, and in this case there is no evident power-law singularity on the log-log plot. Rather, the indication is that $J(q^2)$ also reaches a plateau. Putting these results together, one can calculate a renormalization-group invariant running coupling associated with the ghost-gluon vertex

$$\alpha_s(q^2) = \frac{g^2}{4\pi} J^2(q^2) Z(q^2) \,, \tag{11.13}$$

and this coupling is seen in Fig. 11.6 to run to zero at small momentum. The results displayed in all three of these figures clearly contradict the predictions of the scaling solution.

Fig. 11.4 The lattice gluon propagator in momentum space, $D(q)$ vs. q^2, at $\beta = 5.7$ and lattice volumes up to 96^4. From Bogolubsky et al. [7]

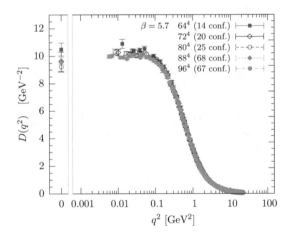

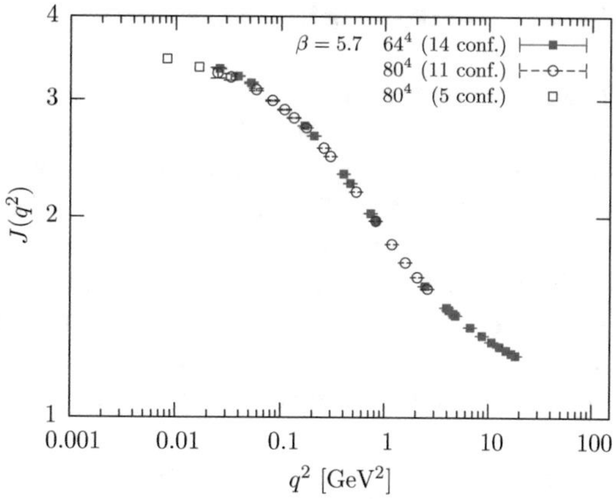

Fig. 11.5 The ghost dressing function in momentum space, $J(q)$ vs. q^2, at $\beta = 5.7$ and lattice volumes of 64^4 and 80^4. From Bogolubsky et al. [7]

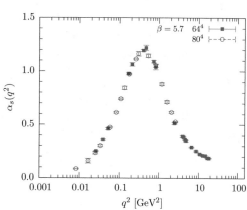

Fig. 11.6 The running coupling $\alpha_s(q^2)$ associated with the ghost-gluon vertex. From Bogolubsky et al. [7]

To make matters a little more confusing, it turns out that the situation is different in $D = 2$ dimensions in Landau gauge. Lattice Monte Carlo simulations have shown that in $D = 2$ dimensions the Landau gauge gluon propagator *does* vanish, and the ghost dressing function *is* singular, in the $q^2 \to 0$ limit. In $D = 2$ dimensions the scaling solution predicts, for the exponents of the ghost and gluon dressing functions,

$$\kappa_{gluon} = 2\kappa_{ghost} + 1 \quad \text{and} \quad \kappa_{ghost} = 1/5 , \qquad (11.14)$$

and these predictions appear to be quite consistent with the $D = 2$ lattice Monte Carlo data obtained by Maas [5].

11.3   The Decoupling Solution

The discouraging results of the numerical simulations in $D > 2$ dimensions have led to a re-examination of the Dyson-Schwinger equations. If the lattice Monte Carlo data is not misleading us, then the scaling solution cannot be the only consistent solution of the Dyson-Schwinger equations. There must be another solution, more in line with the lattice results.

Such a solution was soon found [8], and it is known as the *decoupling solution*. The term "decoupling" refers to the fact that the exponent κ_{ghost} of the ghost dressing function is no longer tied to the exponent κ_{gluon} of the gluon dressing function, as expressed in Eqs. (11.5). Instead, the decoupling solution has

$$J(p^2) \sim \text{const.} \quad , \quad Z(p^2) \sim p^2 \; , \tag{11.15}$$

or $\kappa_{ghost} = 0$, $\kappa_{gluon} = -1$ in the infrared, which clearly violates (11.5). This is not a very exotic solution; the condition on the ghost dressing function just says that the ghost propagator behaves as a simple $1/p^2$ pole near $p = 0$, while $Z(p^2) \sim p^2$ means that the gluon propagator is finite at $p = 0$, as is the case for any massive propagator. To see what is happening, we have to go back to the Dyson-Schwinger equation for the ghost propagator, with couplings and renormalization constants defined with respect to some renormalization point μ. The equation displayed diagrammatically in Fig. 11.1 is, in its full glory [13],

$$J^{-1}(p^2, \mu^2) = \tilde{Z}_3 - \tilde{Z}_1 \frac{g^2(\mu^2)N}{(2\pi)^4} \int d^4 q \; \Gamma_\mu^{(2,1)(0)}(p, q; \mu) \left(\delta_{\mu\nu} - \frac{k_\mu k_\nu}{k^2} \right)$$

$$\times \frac{Z(k^2; \mu^2) J(q^2; \mu^2)}{k^2 q^2} \Gamma_\nu^{(2,1)}(q, p; \mu) \; . \tag{11.16}$$

In this equation, $k \equiv p + q$, $\tilde{Z}_3$ and $\tilde{Z}_1$ are ghost and ghost-gluon vertex renormalization constants, $\Gamma_\mu^{(2,1)(0)}$ is the bare ghost-gluon vertex, and $\Gamma_\mu^{(2,1)}$ is the dressed ghost-gluon vertex. Schematically,

$$J^{-1}(p^2, \mu^2) = \tilde{Z}_3 - \tilde{Z}_1 g^2(\mu^2) I(p; \mu) \; . \tag{11.17}$$

If the ghost dressing function is infrared singular, so that the left hand side of this equation vanishes at $p = 0$, then it must be that the two terms on the right hand side exactly cancel at $p^2 = 0$. But a low-momentum behavior of this type is not the only possibility. One can obtain perfectly valid solutions of the Dyson-Schwinger equations, and these are the decoupling solutions, in which there is not perfect cancellation on the right hand side at $p^2 = 0$ [8]. Another way to put it is that the choice of $J^{-1}(0, \mu^2)$ is a boundary condition for the solution of the Dyson-Schwinger equations [13]. If we subtract, from (11.17), the same equation at $p^2 = 0$, then we have

$$J^{-1}(p^2, \mu^2) = J^{-1}(0, \mu^2) - \tilde{Z}_1 g^2(\mu^2)(I(p; \mu) - I(0; \mu)) . \tag{11.18}$$

The choice $J^{-1}(0, \mu^2) = 0$ generates the scaling solution, and $J^{-1}(0, \mu^2) > 0$ gives the decoupling solution. From this point of view, the scaling solution is only the endpoint of a family of solutions of the Dyson-Schwinger equations.

If the decoupling solution is the correct choice in $D > 2$ dimensions, it means that the Kugo-Ojima criterion cannot be correct (as already argued in Chap. 3), and it also means that we cannot hope to understand the confining static quark potential from a few combinations of vertices and propagators in a covariant gauge. The decoupling solution does, however, allow for the possibility that the gluon propagator violates positivity, so that one doesn't expect to see color charged gluons showing up in the asymptotic spectrum. This is a condition of color confinement, which, as explained in Chap. 3, also holds in a Higgs theory, and so it is a little different from the phenomenon of confinement as it is understood in this book.

As of this writing the decoupling solution seems to be generally accepted. There remains considerable ongoing effort in the Dyson-Schwinger approach which, combined with Bethe-Salpeter equations, is believed by many to be a promising direction for the analytical computation of hadron masses. A good reference is [14].

11.4   Dyson-Schwinger Equations and Coulomb Confinement

Cooper and Zwanziger in [15] have reported some progress towards deriving a confining static quark potential in Coulomb gauge via a Dyson-Schwinger approach. The hypothesis is that the relevant propagators in the infrared limit are instantaneous, i.e. the $D(x - y)$ propagators all contain a factor of $\delta(x_0 - y_0)$, so the propagators are non-zero only at equal times. It is assumed that the ghost propagator $D_{c\bar{c}}$, the propagator D_{AA} for the spacelike components of the gauge field, and the propagator $D_{A_0 A_0}$ for the timelike components have the momentum space power law behavior in $k = \mathbf{k}$ at low spacelike momenta

$$D_{c\bar{c}} \propto \frac{1}{k^\alpha} \quad , \quad D_{AA} \propto \frac{1}{k^\gamma} \quad , \quad D_{A_0 A_0} \propto \frac{1}{k^\delta} \tag{11.19}$$

This power behavior, when inserted into a set of Coulomb gauge Dyson-Schwinger equations, yields a set of consistency equations for the critical indices α, γ, δ which, it is found, can only be satisfied below a critical space dimension

$$d_c = 2.9677 \tag{11.20}$$

Table 11.1 Comparison of Schwinger-Dyson equations to Lattice, from Cooper and Zwanziger [15]

Critical exponents		LM notation		SDE at d_c		Lattice calculation
$\alpha = (d+2)/2$	$=$	$2\kappa + 2$	$=$	2.4839	$\approx$	2.490(10)
$\gamma = 0$			$=$	0	$\approx$	0
$\delta = d + 1 + \theta_{\pm}(d)$	$=$	$2\delta_{LM} + 2$	$=$	3.9677	$\approx$	4.10(10)

which is only slightly below the desired limit (for a $D = 3 + 1$ dimensional theory) at $d = 3$. Evaluating the critical indices at this critical dimension yields the results shown in Table 11.1, which are compared to the results for the same indices obtained from lattice simulations due to Langfeld and Moyaerts (LM) [16]. Clearly the agreement is very good. A critical index $\delta = 4$ corresponds to a linearly rising color Coulomb potential, and the index in the table below is very close to that. One caveat, however, is that in $d = 2$ space dimensions this Dyson-Schwinger approach results in a color Coulomb potential rising much faster than linear, $V(r) \propto r^2$.

11.5 Effective Polyakov Line Potential

Another direction, pursued by Pawlowski and collaborators [17, 18], is to derive an effective potential $V(A_0)$ which would allow one to show that the expectation value of a Polyakov line vanishes at low temperatures, is non-zero at high temperatures, with a first order or continuous phase transition, depending on the gauge group, separating the two regimes. Most of this work is done in an approach known as the Functional Renormalization Group (FRG) invented by Wetterich [19]. The general idea is along the lines of the Wilson "block spin" approach; i.e. one seeks to obtain effective actions at some scale, having integrated out fluctuations at momenta higher than that scale. Of course a practical implementation of this scheme involves certain truncations. We will not try to explain this approach here (the reader is referred to the cited references), except to note that in practice it has strong similarities to the Dyson-Schwinger approach. In this section we will only display some of the main results obtained by these methods, as they concern the confinement problem.

The Polyakov loop for SU(N) gauge theory in the continuum is defined, with periodic boundary conditions in the time direction, as

$$P(\mathbf{x}) = \frac{1}{N} \text{Tr} \mathcal{P} \left\{ \exp \left[ig \int_0^\beta dt\, A_0(\mathbf{x}, t) \right] \right\} \qquad (11.21)$$

where $\mathscr{P}$ denotes path-ordering in time, and $\beta = 1/kT$ where T is temperature. It is possible to fix to a Polyakov line gauge in which $A_0(\mathbf{x}, t) = A_0(\mathbf{x})$ is time independent, and moreover lies in the Cartan subalgebra of the SU(N) Lie algebra. In this gauge, for the SU(2) group, we may write

$$\beta g A_0(\mathbf{x}) = 2\pi \varphi(\mathbf{x}) \tfrac{1}{2}\sigma_3$$

$$P(\mathbf{x}) = \frac{1}{2}\text{Tr}\{\exp[ig\beta A_0(\mathbf{x})]\}$$

$$= \cos(\pi\varphi(\mathbf{x})) \qquad (11.22)$$

Likewise, for the SU(3) group,

$$\beta g A_0(\mathbf{x}) = 2\pi(\varphi_3(\mathbf{x})T^3 + \varphi_8(\mathbf{x})T^8) \qquad (11.23)$$

where T^3, T^8 are the generators in the Cartan subalgebra, so that

$$P(\mathbf{x}) = \frac{1}{3}\left(e^{-2\pi i \varphi_8(\mathbf{x})/\sqrt{3}} + 2\cos(\pi\varphi_3(\mathbf{x}))e^{\pi i \varphi_8(\mathbf{x})/\sqrt{3}}\right) \qquad (11.24)$$

The FRG approach is then used to compute the effective potential $V(A_0)$, with $\langle A_0 \rangle$ obtained at the minimum of this potential. It is argued in [17] that $P[\langle A_0 \rangle] = 0$ implies that $\langle P \rangle = 0$. The results for this effective potential, for several temperatures near the deconfinement temperature, are displayed in Figs. 11.7a and b for the SU(2) and SU(3) groups. The x-axis in both subfigures $\varphi = \varphi_3$, and in the SU(3) case this is at $\varphi_8 = 0$. What is clear from both figures is that at the lower temperatures, $P[\langle A_0 \rangle] = 0$, while at higher temperatures $P[\langle A_0 \rangle] \neq 0$. Moreover, inspection of these figures suggests a first order transition in SU(3), and a continuous transition in SU(2), with transition temperatures that appear to be compatible with the lattice results.

These are very encouraging results with respect to Polyakov lines. Deriving an area law for Wilson loops is another matter, and efforts using propagators and vertices derived from Dyson-Schwinger (decoupling solution) and FRG methods have so far arrived only at a perimeter law.

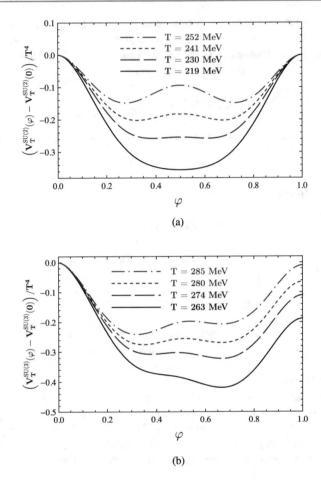

Fig. 11.7 The effective potential $V(A_0)$, derived in the Functional Renormalization Group approach, for (**a**) SU(2) and (**b**) SU(3) gauge theories, at a number of temperatures in the vicinity of the deconfinement transition. Note that below the transition temperatures the minima of these effective potentials occur at points for which $P[A_0] = 0$. For the relation of φ to A_0, see the text. From [18]

References

1. L. von Smekal, R. Alkofer, A. Hauck, The infrared behavior of gluon and ghost propagators in Landau gauge QCD. Phys. Rev. Lett. **79**, 3591 (1997), hep-ph/9705242; A solution to coupled Dyson-Schwinger equations for gluons and ghosts in Landau gauge. Ann. Phys. **267**, 1 (1998) [Erratum-ibid. **269**, 182 (1998)], hep-ph/9707327; C.S. Fischer, R. Alkofer, H. Reinhardt, Phys. Rev. D **65**, 094008 (2002), hep-ph/0202195; C.S. Fischer, R. Alkofer, The elusiveness of critical exponents in Landau gauge Yang-Mills theories. Phys. Lett. B **536**, 177 (2002), hep-ph/0202202
2. C.S. Fischer, Infrared properties of QCD from Dyson-Schwinger equations. J. Phys. G **32**, R253 (2006). arXiv:hep-ph/0605173

3. T. Kugo, The universal renormalization factors Z(1) / Z(3) and color confinement condition in non-Abelian gauge theory (1995). arXiv:hep-th/9511033
4. D. Zwanziger, Vanishing of zero momentum lattice gluon propagator and color confinement. Nucl. Phys. B **364**, 127 (1991); Renormalizability of the critical limit of lattice gauge theory by BRS invariance. Nucl. Phys. B **399**, 477 (1993)
5. A. Maas, Two- and three-point Green's functions in two-dimensional Landau-gauge Yang-Mills theory. Phys. Rev. D **75**, 116004 (2007). arXiv:0704.0722 [hep-lat]
6. A. Cucchieri, T. Mendes, What's up with IR gluon and ghost propagators in Landau gauge? A puzzling answer from huge lattices. PoS **LAT2007**, 297 (2007). arXiv:0710.0412 [hep-lat]; A. Cucchieri, T. Mendes, Constraints on the IR behavior of the ghost propagator in Yang-Mills theories. Phys. Rev. D **78**, 094503 (2008). arXiv:0804.2371 [hep-lat]; Constraints on the IR behavior of the gluon propagator in Yang-Mills theories. Phys. Rev. Lett. **100**, 241601 (2008). arXiv:0712.3517 [hep-lat]
7. I.L. Bogolubsky, E.M. Ilgenfritz, M. Muller-Preussker, A. Sternbeck, Lattice gluodynamics computation of Landau gauge Green's functions in the deep infrared. Phys. Lett. B **676**, 69 (2009). arXiv:0901.0736 [hep-lat]
8. Ph. Boucaud, J.P. Leroy, A.L. Yaouanc, J. Micheli, O. Pene, J. Rodriguez-Quintero, IR finiteness of the ghost dressing function from numerical resolution of the ghost SD equation. JHEP **0806**, 012 (2008). arXiv:0801.2721 [hep-ph]; A.C. Aguilar, D. Binosi, J. Papavassiliou, Gluon and ghost propagators in the Landau gauge: Deriving lattice results from Schwinger-Dyson equations. Phys. Rev. D **78**, 025010 (2008). arXiv:0802.1870 [hep-ph]; D. Dudal, J.A. Gracey, S.P. Sorella, N. Vandersickel, H. Verschelde, A refinement of the Gribov-Zwanziger approach in the Landau gauge: infrared propagators in harmony with the lattice results. Phys. Rev. D **78**, 065047 (2008). arXiv:0806.4348 [hep-th]
9. C. Itzykson, J.-B. Zuber, *Quantum Field Theory* (McGraw-Hill, New York, 1980)
10. C.S. Fischer, Non-perturbative propagators, running coupling and dynamical mass generation in ghost - antighost symmetric gauges in QCD (2003). arXiv:hep-ph/0304233
11. C. Lerche, L. von Smekal, On the infrared exponent for gluon and ghost propagation in Landau gauge QCD. Phys. Rev. D **65**, 125006 (2002). arXiv:hep-ph/0202194; D. Zwanziger, Non-perturbative Landau gauge and infrared critical exponents in QCD. Phys. Rev. D **65**, 094039 (2002). arXiv:hep-th/0109224
12. R. Alkofer, C.S. Fischer, F.J. Llanes-Estrada, K. Schwenzer, The quark-gluon vertex in Landau gauge QCD: Its role in dynamical chiral symmetry breaking and quark confinement. Ann. Phys. **324**, 106 (2009). arXiv:0804.3042 [hep-ph]
13. C.S. Fischer, A. Maas, J.M. Pawlowski, On the infrared behavior of Landau gauge Yang-Mills theory. Ann. Phys. **324**, 2408 (2009). arXiv:0810.1987 [hep-ph]
14. G. Eichmann, Hadron phenomenology in the Dyson-Schwinger approach. J. Phys. Conf. Ser. **426**, 012014 (2013)
15. P. Cooper, D. Zwanziger, Schwinger-Dyson equations in coulomb gauge consistent with numerical simulation. Phys. Rev. D **98**(11), 114006 (2018). arXiv:1803.06597 [hep-th]
16. K. Langfeld, L. Moyaerts, Propagators in coulomb gauge from SU(2) lattice gauge theory. Phys. Rev. D **70**, 074507 (2004). hep-lat/0406024
17. F. Marhauser, J.M. Pawlowski, Confinement in Polyakov gauge (2008). arXiv:0812.1144 [hep-ph]
18. L. Fister, J.M. Pawlowski, Confinement from correlation functions. Phys. Rev. D **88**, 045010 (2013). arXiv:1301.4163 [hep-ph]
19. C. Wetterich, Exact evolution equation for the effective potential. Phys. Lett. B **301**, 90 (1993). https://doi.org/10.1016/0370-2693(93)90726-X. arXiv:1710.05815 [hep-th]

Large-N, Planar Diagrams, and the Gluon-Chain Model

12

Abstract

Organization of Feynman diagrams in an SU(N) gauge theory as double expansion in 't Hooft coupling $\lambda = g^2 N$, and in $1/N_{colors}^2$. The factorization property at large N, and Casimir scaling. The gluon chain model.

The Lagrangian of a pure $SU(N)$ gauge theory

$$S = \int d^4x \, \frac{1}{2} \text{Tr}[F_{\mu\nu} F_{\mu\nu}]$$

$$F_{\mu\nu} = \partial_\mu A_\nu - \partial_\nu A_\mu - ig[A_\mu, A_\nu] \tag{12.1}$$

appears to contain only one parameter, namely the coupling g, and one would think that the Feynman diagrammatic expansion for Green's functions, n-point correlation functions, Wilson loops, and other observables can only be organized in powers of g. But it was pointed out by 't Hooft [1], in 1974, that the number of colors N of the $SU(N)$ gauge group can also be regarded as a parameter, and the Feynman diagram series can be reorganized as a double expansion in powers of $1/N^2$, and in powers of the *'t Hooft coupling*

$$\lambda \equiv g^2 N \,, \tag{12.2}$$

so that the diagrammatic series for any observable has the form

$$\langle O \rangle = N^p \sum_{n=0}^{\infty} N^{-2n} T_n(\lambda) \,, \tag{12.3}$$

© Springer Nature Switzerland AG 2020
J. Greensite, *An Introduction to the Confinement Problem*, Lecture Notes in Physics 972, https://doi.org/10.1007/978-3-030-51563-8_12

where each of the $T_n(\lambda)$ is an (infinite) series in powers of λ. If we take the $N \to \infty$ limit with the 't Hooft coupling λ held fixed, then only the leading term T_0 is important, and this term can usually be represented by *planar diagrams*, which are Feynman diagrams that can be drawn on a sheet of paper, such that no gluon line crosses over any other gluon line.

The planar contribution seems to dominate in $SU(N)$ theories for many observables, even down to $N = 2$. Lattice calculations for, e.g., the string tension σ, from $N = 2$ to $N = 5$, indicate that the tension is almost independent of N when the 't Hooft coupling λ is held fixed, while the dimensionless ratio of glueball mass m to $\sqrt{\sigma}$ varies only a little as N is increased from 2 to 5 [2]. The great hope is that certain remarkable simplifications, that are inherent in the $N = \infty$ limit, may provide new insights into (or perhaps even allow a solution of) non-abelian gauge theories in this limit.

12.1 Double-Line Notation and Factorization

The rationale for the double expansion in λ and $1/N$ begins with the observation that the gluon field is a matrix-valued field, i.e. $A_\mu^{ab} = A_\mu^c L_c^{ab}$, where the $\{L_c\}$ are $SU(N)$ group generators, and a, b are matrix (or "color") indices which run from 1 to N. The gluon propagator is then

$$\langle A_\mu^{ab}(x) A_\nu^{cd}(y)\rangle_0 = (\delta^{ad}\delta^{bc} - \frac{1}{N}\delta^{ab}\delta^{cd}) D_{\mu\nu}(x - y) . \tag{12.4}$$

When N is very large, the term proportional to $1/N$ on the right hand side is unimportant, and can be dropped. Alternatively, we may consider the gauge group $U(N)$ as an approximation to $SU(N)$. This gauge group is essentially a product $U(1) \times SU(N)$, since one gluon decouples from the others, and for most gauge-invariant observables that single gluon, out N^2 gluons, makes a negligible contribution when N is very large. For the $U(N)$ gauge group the $1/N$ term in (12.4) is absent from the start, and the color indices at the beginning of a gluon propagator match the indices at the end. This fact motivates 't Hooft's double-line notation for the gluon propagator, in which each line carries a matrix index. Then, because the interaction terms are just products of matrices (i.e. matrix-valued fields), there is a simple flow of color indices through the interaction vertices which can also be represented in double line notation, as shown in Fig. 12.1. Using this notation, it is quite straightforward to count the power of N associated with any particular diagram. The general rule is that any single line which forms a closed loop contributes a factor of N, because it represents a color index which is summed from 1 to N.

As an important example, let us consider the perturbative evaluation of a Wilson loop, assuming that the loop is small enough to be amenable (because of asymptotic freedom) to such methods. Figure 12.2 shows a few low-order diagrams in both ordinary and double-line notation, and for purposes of index counting the dashed

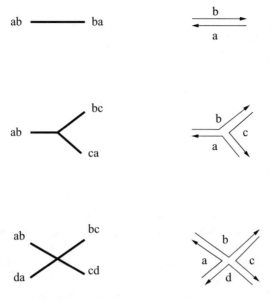

Fig. 12.1 't Hooft double-line notation for propagators and vertices. Only the index dependence is shown; vector indices are suppressed

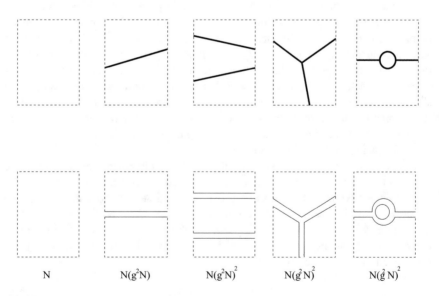

Fig. 12.2 Planar Feynman diagram contributions to a rectangular Wilson loop, in the usual (upper row) and double-line (lower row) notation

Fig. 12.3 The leading
non-planar contribution to a
rectangular Wilson loop, due
to two-gluon exchange, in the
usual and in double-line
notation

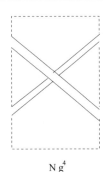

$$N g^4$$

lines representing the Wilson loop are to be counted as part of the index loops. The first, zeroth-order diagram is just equal to the number N of colors. The second diagram, with one gluon exchange, has two powers of the gauge coupling, and two index loops which contribute a factor of N^2. So the diagram is of order $g^2 N^2$, which we write as $N\lambda$. The next three diagrams are fourth order in the coupling, but have three index loops, and are therefore of order $g^4 N^3$, equal to $N\lambda^2$. The pattern (an overall factor of N, times a power of λ) is becoming clear, and it can be shown that the contribution of any planar diagram (no crossing gluon lines) is of order $N\lambda^p$. These diagrams contribute to the term $NT_0(\lambda)$ in the series (12.3).

On the other hand, consider the leading non-planar contribution shown in Fig. 12.3. This diagram is fourth-order in the coupling, but has only a single index loop, so it is of order $g^4 N$ which can be expressed as $(1/N^2)(N\lambda^2)$. That is the leading contribution to the term NT_1 in the large-N series expansion of the Wilson loop. In general if we take the simultaneous limits $N \to \infty$ and $g \to 0$ in such a way that the 't Hooft coupling $\lambda = g^2 N$ is non-zero and finite in the limit, then all non-planar contributions are negligible, i.e. down by a power of at least $1/N^2$ as compared to the planar contribution.[1] A similar reorganization can be carried out in the strong-coupling expansion of lattice gauge theory; in that case the double expansion is in powers of $1/N^2$ and $1/\lambda$.

If it would be possible to sum up all the planar Feynman diagrams for large Wilson loops at $N = \infty$, and demonstrate confinement in that limit, that would be a spectacular advance. This has so far proved impossible. The hope that it *might* be possible is based in part on a striking property of the $N = \infty$ limit which seems quite unlike the case for $N = 2$ or $N = 3$, and this is the property of *factorization*. Let O_1 and O_2 be any two gauge-invariant operators. The factorization property says that, to leading order in $1/N^2$,

$$\langle O_1 O_2 \rangle = \langle O_1 \rangle \langle O_2 \rangle . \tag{12.5}$$

---

[1] The power of N associated with any graph actually depends on certain topological properties of the graph, cf. [1] and [3].

For example, let O_1 and O_2 be Wilson loops. Feynman diagrams which spoil the equality of the left and right hand sides of (12.5) are diagrams in which gluon lines connect the two loops, and these diagrams are always down by (at least) a factor of $1/N^2$, compared to planar diagrams in which there are no gluon lines connecting the loops.

An immediate consequence of the factorization property is Casimir scaling. Let $U(C)$ be a Wilson loop holonomy. Then the trace of the holonomy in group representation r (the group character $\chi_r[U(C)]$) can be always be expressed in terms of products of characters in the fundamental defining representation. To leading order in $1/N^2$

$$\chi_r[U(C)] = \left(\chi_F[U(C)]\right)^n \left(\chi_F^*[U(C)]\right)^{\bar{n}} + \text{sub-leading terms} . \tag{12.6}$$

Taking the expectation value of both sides, and applying the factorization property (12.5), it follows that as $N \to \infty$ the string tension σ_r of loops in representation r is an integer multiple of the string tension σ_F in the fundamental representation, i.e.

$$\sigma_r = (n + \bar{n})\sigma_F . \tag{12.7}$$

In particular, the string tension of the adjoint loop is simply twice the string tension of the fundamental loop $\sigma_{adj} = 2\sigma_{fund}$, Now in the large-N limit, the quadratic Casimir of representation r is $C_r = (n + \bar{n})N/2$. Then (12.7) is simply the statement of Casimir scaling.

In the $N = \infty$ limit, the Casimir scaling regime extends from the confinement scale out to infinity, and this is in contrast to any theory at finite N, where the asymptotic string tensions depend only on N-ality. The reason for this difference is that string breaking is a $1/N^2$-suppressed process [4], and therefore the loop size where string-breaking occurs, and the asymptotic string tension is obtained, runs off to infinity as $N \to \infty$.[2]

The factorization property also implies, since $\langle O^2 \rangle = \langle O \rangle^2$ and by definition $\Delta O^2 = \langle O^2 \rangle - \langle O \rangle^2$, that the rms deviation ΔO of any gauge-invariant observable from its mean value vanishes in the large-N limit. This fact has an astonishing consequence. If we would imagine carrying out a numerical simulation of $SU(N)$ gauge theory at some enormous value of N, such that all non-leading powers of N could be neglected, then a gauge-invariant observable, evaluated in any set of thermalized configurations, would have the same value in each configuration. This means that it is not necessary to average over many configurations to obtain the expectation value of a gauge-invariant observable; a single configuration (if it is the right configuration) would be sufficient. This led Witten [6] to the idea of the large-N *master field*; i.e. there should exist a field configuration A_μ^{master} such that for any

---

[2]For a discussion of the center vortex theory at large but finite N, cf. [5].

gauge invariant observable O, up to $1/N^2$ corrections,

$$\langle O \rangle = O[A_\mu^{master}] \,, \tag{12.8}$$

where the right hand side is simply the operator evaluated at the master field configuration. This is very different from what happens at finite N. For example, In a Monte Carlo evaluation of a large Wilson loop $\text{Tr}[U(C)]/N$ in $SU(2)$ or $SU(3)$ gauge theory, the trace of the loop is usually not small, when evaluated in a typical thermalized configuration, and the trace divided by N varies wildly between ± 1. The expectation value of this quantity becomes exponentially small, for large loops, due to near-perfect cancellations between the large positive and negative values obtained in different configurations. In contrast, as $N \to \infty$, the same quantity is tiny, for large loops, in *any* thermalized configuration. For loops evaluated in such configurations, or in the master field configuration, the wild fluctuations between positive and negative values, and the near-perfect cancellation, really takes place among the eigenvalues of the loop holonomy $U(C)$, which nearly cancel among themselves when taking the trace.[3]

It would be wonderful to find a master field for $N = \infty$ gauge theory which has such marvelous properties. Unfortunately, although the master field is known explicitly for certain models with matrix-valued fields in very low dimensions [8], where knowing the master field amounts to knowing its eigenvalue spectrum, we do not know the master field for $U(\infty)$ gauge theories in the interesting cases of $D = 3$ and $D = 4$ dimensions. Despite this fact, one has the feeling that the large-N limit is still a huge simplification in non-abelian gauge theory, and a number of important advances in the field are based on it. Perhaps first among these is the AdS/CFT correspondence, to be discussed in Chap. 14. Another intriguing idea is known as *large-N reduction*, in which the D-dimensional theory is replaced by a zero-dimensional theory; i.e. a theory defined at a single point. The original proposal, due to Eguchi and Kawai [9] was to replace the periodic lattice in D-dimensional lattice gauge theory by a single (periodic) hypercube, and this turns out to work at strong, but not at weak couplings. More sophisticated versions of large-N reduction were introduced soon after [10, 11], but it has been shown recently that these models also encounter technical difficulties at weak couplings [12], at least for pure gauge theories. In any case, it does not seem to be any easier analytically to solve the reduced models than to solve the full theory.

A further feature of the large-N expansion, pointed out by 't Hooft in his seminal work [1], is more suggestive than quantitative. Suppose that the diagrammatic expansion would somehow be relevant to large Wilson loops, and that the expansion is dominated by very high order planar diagrams, as indicated schematically (in double-line notation) in Fig. 12.4. This high-order diagram *looks* like a discretized surface, and if we would take a time-slice, the particle content is a static quark and

---

[3]In $D = 2$ dimensions and $N = \infty$, the spectral density of the eigenvalues of $U(C)$ was worked out by Durhuus and Olesen [7].

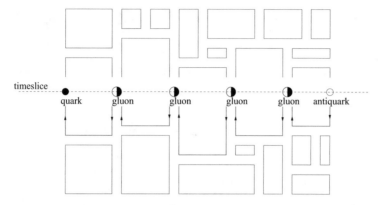

Fig. 12.4 The gluon chain as a time slice of a planar diagram, which is shown here in double-line notation for propagators introduced by 't Hooft. A solid hemisphere indicates a quark color index, and an open hemisphere an antiquark color index

antiquark source, with a series of virtual gluons between the sources. This picture suggests a discretized string of some kind, leading to the idea that the color-electric flux tube between a quark-antiquark pair might be thought of as a chain of gluons, bound together by attractive forces. This is known as the *gluon-chain model* (cf. [13] and references therein, and ref. [14]), and it has a number of attractive features. In the next section we will discuss some tests of this idea, as formulated in Coulomb gauge.

12.2   The Gluon Chain Model

We have seen in Chap. 10 that the color Coulomb potential is linear in pure $SU(2)$ gauge theory, but the corresponding Coulomb string tension is about four times larger than the asymptotic string tension of the static quark potential. This raises the question, given the existence of a linear Coulomb potential, of how the string tension is reduced, and how a flux tube forms in Coulomb gauge.

Let us recall that the Coulomb potential is simply the R-dependent part of the energy expectation value in the quark-antiquark physical state

$$\Psi_{q\bar{q}} = \overline{q}^a(0)q^a(R)\Psi_0 . \tag{12.9}$$

But there is no reason that this state should be the minimal energy state out of all states containing a static quark-antiquark pair of separation R, and it should be possible to construct lower-energy states with the help of additional gluon operators, creating "constituent" gluons. Schematically, we are looking for lower energy states

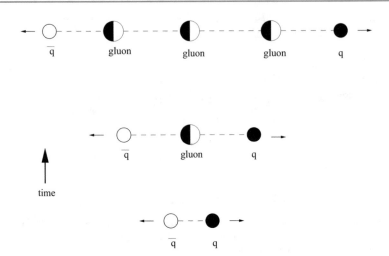

Fig. 12.5 Formation of a gluon chain, as a quark-antiquark pair separates

of the form

$$\Psi'_{q\bar{q}} = \bar{q}^a(0)\left\{ c_0\delta^{ab} + c_1 A^{ab} + c_2 A^{ac} A^{cb} + \ldots \right\} q^b(R)\Psi_0 . \qquad (12.10)$$

A Wilson line for example, running between the quark-antiquark pair has exactly this form, if one expands the path-ordered exponential in powers of the A_μ-field, and has no color electric field whatever off the line itself.

The gluon-chain model is the idea that a quark-antiquark pair, as they separate, pull out a chain of gluons (Fig. 12.5), with nearest neighbors in the chain bound by attractive forces between them. Thus the minimal energy state $\Psi'_{q\bar{q}}$ will be dominated, for large quark-antiquark separations, by components with many constituent gluons. In its original form, it was supposed that as the quark and antiquark separate, the field energy increases faster than linearly, and at some point it is energetically favorable to insert a gluon between the quark-antiquark charges to reduce the effective charge separation. In fact, we have seen that the Coulomb energy grows only linearly, but it may still be the case that the energy of physical states with widely separated color charges can be lowered by the inclusion of constituent gluons. This question has been studied, first in the context of a simple model [15], and later more quantitatively in [16]. The idea is to begin from a set of orthogonal trial states, denoted by $|n\rangle$, with static quark and antiquark separated by R, and each state containing a given number of gluon operators acting on the vacuum. One then computes the Coulomb gauge Hamiltonian matrix elements H_{mn} between these states, and his can be done quasi-perturbatively, using ghost, gluon, and Coulomb propagators derived from the lattice. The Hamiltonian matrix is then diagonalized, and the lowest eigenvalue is minimized with respect to a set of variational parameters. The final result is a variational estimate of the static quark

Fig. 12.6 Adjoint string-breaking in the gluon chain model. Two gluons in separate chains (I) scatter by a contact interaction, resulting in the re-arrangement of color indices indicated in II. This corresponds to chains starting and ending on the same heavy source. The chains then contract down to smaller "gluelumps" (III)

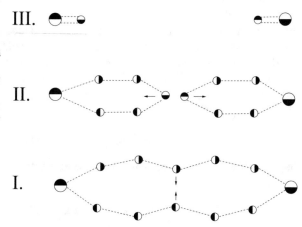

potential $V(R)$. In this calculation it is found that the string tension can indeed be much less than the Coulomb string tension [16].

One of the strongest motivations for the gluon chain model is that a gluon chain can be regarded as a time-slice of a high-order planar Feynman diagram, as we have already seen in Fig. 12.4. The gluon chain also has string-like properties (e.g. a Lüscher term) due to fluctuations in the position of its gluon constituents, it features Casimir scaling at large N, and also has the right N-ality properties due to string-breaking [13]. For example, at large N there are two chains between heavy sources in the adjoint representation, giving rise to twice the string tension as in the fundamental representation, which is the correct Casimir ratio at large N. Interaction between the two chains is $1/N^2$ suppressed, but at finite N the chains can interact and rearrange themselves (Fig. 12.6) into two "gluelumps," having negligible energy dependence on the quark separation R. In general, for a quark and antiquark in representation r satisfying (12.6), there will be $n + \bar{n}$ chains running between the static sources. Neglecting string-breaking or other interactions between the chains, this leads to the large-N Casimir scaling result $\sigma_r = (n + \bar{n})\sigma_F$, and when string-breaking is taken into account, the number of remaining chains depends only on the N-ality of representation r.

12.2.1 Numerical Investigation in Coulomb Gauge

In Euclidean lattice gauge theory, the transfer matrix $\mathscr{T} = \exp[-Ha]$ is the Euclidean time version of the Minkowski space time evolution operator, where a is lattice spacing in the time direction. It is useful to define the rescaled transfer matrix

$$T = \exp[-(H - E_0)a] , \tag{12.11}$$

where E_0 is the vacuum energy. To compute the static quark potential of a quark-antiquark pair separated by a distance R, one would ideally diagonalize the transfer matrix in the infinite-dimensional subspace of states which contain a single massive quark, and a single massive antiquark, located at sites $\mathbf{x}$ and $\mathbf{y}$ with $R = |\mathbf{x} - \mathbf{y}|$. The minimal energy eigenstate of the transfer matrix, in this subspace, is the state with the largest eigenvalue λ_{max} of T, and the energy of the state in lattice units is given by

$$V(R) = -\log(\lambda_{max}) . \tag{12.12}$$

In practice, we seek to diagonalize T in the subspace spanned by a finite number of $q\bar{q}$ states of the form

$$|k\rangle = \bar{q}^a(\mathbf{x}) Q_k^{ab} q^b(\mathbf{y}) |\Psi_0\rangle \quad , \quad k = 1, 2, .., M , \tag{12.13}$$

where the Q_k operators are functionals of the lattice link variables. This program has recently been carried out in ref. [17], with a set of $M = 6$ operators involving up to two gluon fields on the lattice, creating up to two "constituent" gluons in the region between the quark and antiquark sources. The Q_k operators specify the set of $\{|k\rangle\}$ states, and from these an orthonormal set $\{|\varphi_k\rangle\}$ is constructed by the Gram-Schmidt procedure. The transfer matrix is diagonalized in the subspace spanned by the $\{|\varphi_k\rangle\}$, and the lowest energy eigenstate in this subspace

$$|\psi(R)\rangle = \sum_{k=1}^{6} a_k(R)|\varphi_k\rangle \tag{12.14}$$

is determined. The corresponding energy depends on a variational parameter in the Q_k operators, which is then selected to minimize the energy of the lowest energy state. Since φ_1 is just the vacuum (no gluon) state, φ_2 is a one-gluon state, and the remainder are two-gluon states, it follows that a_1^2 is the fraction of the norm of $\psi(R)$ due to the zero (constituent) gluon component $|\varphi_1\rangle = |1\rangle$, a_2^2 is the fraction of the norm $\psi(R)$ from the one-gluon component $|\varphi_2\rangle \propto |2\rangle$, and $1 - a_1^2 - a_2^2$ is the fraction of the norm due to the remaining two-gluon component states.[4] The energy of this state is

$$V_{chain}(R) = -\log(\lambda_{max}) , \tag{12.15}$$

which can be compared to the Coulomb (self+interaction) energy

$$V_C(R) = -\log(T_{11}) , \tag{12.16}$$

where $T_{11} = \langle \varphi_1 | T | \varphi_1 \rangle = \langle 1 | T | 1 \rangle$ is the matrix element of T in the zero constituent gluon state. Note that $V_C(R) = V(R, 0)$, in the notation of Sect. 9.2.

---

[4]For the explicit form of the $\{Q_k\}$ cf. [17].

The resulting potentials, for $SU(2)$ lattice gauge theory at $\beta = 2.4$ are shown in Fig. 12.7, where the static quark potential, computed by standard methods, is also displayed. Two features worth emphasizing in this figure is that $V_{chain}(R)$ remains linear, and its slope is much closer to that of the asymptotic string tension, as compared to the Coulomb string tension.

Figure 12.8 is a comparison of the zero, one, and two constituent-gluon content of the minimal energy state at each R, where by "content" is meant the fraction of

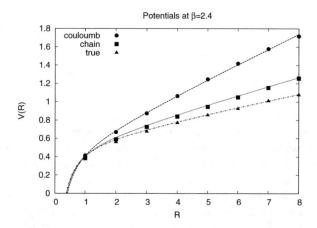

Fig. 12.7 The color Coulomb potential $V_{coul}(R)$, the "gluon-chain" potential $V(R)$ derived from the minimal-energy variational state, and the "true" static quark potential $V_{true}(R)$ obtained by standard methods. Results are shown at lattice coupling $\beta = 2.4$. Continuous lines are from a fit of data points to $\sigma R - \pi/(12R) + c$

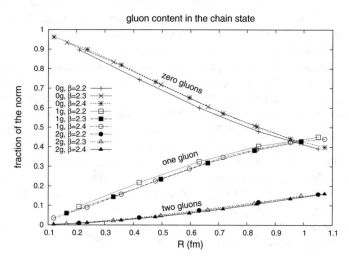

Fig. 12.8 Zero, one, and two-gluon content (fraction of the norm of the variational state) vs. quark separation R in fermis, at $\beta = 2.2, 2.3, 2.4$

Fig. 12.9 Sensitivity of the
Coulomb potential $V_{coul}(R)$
(solid symbols), and
insensitivity of the chain
potential $V_{chain}(R)$ (open
symbols), to lattice volume.
Data is for the gauge coupling
$\beta = 2.4$, and lattice volumes
$L^4 = 12^4, 16^4, 22^4$.
Quark-antiquark separation R
is in lattice units

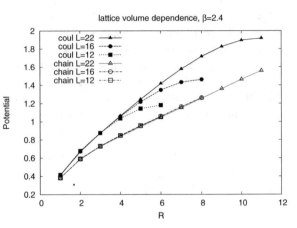

the norm of the minimal energy state. The zero gluon content obviously dominates
at small $q\bar{q}$ separation. This is of course expected, because placing a gluon between
the quark and antiquark comes with a price in kinetic energy. Around $R \approx 1$ fm,
however, the zero and one-gluon contributions to the minimal energy state are about
equal. It is important to note also that the gluon content vs. R in physical units is
almost coupling independent, which serves as a check of the whole procedure.

Figure 12.9 shows the Coulomb energy $V_C(R)$, and the multi-gluon variational
state energy $V_{chain}(R)$ computed at $\beta = 2.4$, at a variety of lattice sizes. The
Coulomb energy shows a strong sensitivity to lattice extension, with the potential
flattening out as $R \to L/2$. In contrast, $V_{chain}(R)$ appears to be rather insensitive
to the lattice size. This suggests that the chain state has no long-range dipole field,
or at least that the long-range field is greatly suppressed relative to that of the zero-
gluon state. It is possible that, in the multi-gluon state, we are beginning to see the
structure of the minimal energy color electric flux tube.

I have already noted that one of the main motivations of the gluon chain model
is the resemblance of high-order planar Feynman diagrams to a discretized surface,
and the time-slices of such diagrams to a discretized string; i.e. a chain. There is
an approach due to Bardakci and Thorn [18], which explicitly formulates the planar
Feynman diagrams of matrix ϕ^3 field theory as the diagrams of a discretized string
theory in light-cone coordinates. This approach has been extended to gauge theories
by Thorn in ref. [19]. Since this is a string formulation of any planar diagram,
including low-order Feynman diagrams, the discretized string formulation is not
sufficient in itself to establish the existence of confinement, the QCD flux tube,
or the gluon-chain nature of the QCD flux tube. On the other hand, the "stringy"
properties of a gluon-chain seem rather natural in this approach; perhaps gluon-
chain excitations will emerge in some new approximation scheme, specifically
adapted to the Bardakci-Thorn reformulation of planar QCD.

References

1. G. 't Hooft, A planar diagram theory for strong interactions. Nucl. Phys. B**72**, 461 (1974)
2. B. Lucini, M. Teper, SU(N) gauge theories in four dimensions: exploring the approach to N = infinity. J. High Energy Phys. **0106**, 050 (2001). [arXiv:hep-lat/0103027]
3. S. Coleman, *Aspects of Symmetry* (Cambridge University Press, Cambridge, 1995)
4. J. Greensite, M.B. Halpern, Suppression of color screening at large N. Phys. Rev. D **27**, 2545 (1983)
5. J. Greensite, S. Olejnik, k-string tensions and center vortices at large N. J. High Energy Phys. **0209**, 039 (2002). [arXiv:hep-lat/0209088]
6. E. Witten, in *Recent Developments in Gauge Theories*, ed by G. 't Hooft. 1979 Cargese Lectures (Plenum, New York, 1980)
7. B. Durhuus, P. Olesen, The spectral density for two-dimensional continuum QCD. Nucl. Phys. B **184**, 461 (1981)
8. E. Brezin, C. Itzykson, G. Parisi, J.B. Zuber, Planar diagrams. Commun. Math. Phys. **59**, 35 (1978)
9. T. Eguchi, H. Kawai, Reduction of dynamical degrees of freedom in the large N Gauge theory. Phys. Rev. Lett. **48**, 1063 (1982)
10. G. Bhanot, U.M. Heller, H. Neuberger, The quenched Eguchi-Kawai model. Phys. Lett. B **113**, 47 (1982)
11. A. Gonzalez-Arroyo, M. Okawa, A twisted model for large N lattice Gauge Theory. Phys. Lett. B **120**, 174 (1983)
12. B. Bringoltz, S.R. Sharpe, Breakdown of large-N reduction in the quenched Eguchi-Kawai model. PoS **LATTICE2008**, 055 (2008). [arXiv:0810.1239 [hep-lat]]
13. J. Greensite, C.B. Thorn, Gluon chain model of the confining force. J. High Energy Phys. **0202**, 014 (2002). [arXiv:hep-ph/0112326]
14. G. Tiktopoulos, Gluon Chains. Phys. Lett. B **66**, 271 (1977)
15. J. Greensite, A.P. Szczepaniak, Coulomb string tension, asymptotic string tension, and the gluon chain. Phys. Rev. D **91**(3), 034503 (2015). https://doi.org/10.1103/PhysRevD.91.034503. [arXiv:1410.3525 [hep-lat]]
16. J. Greensite, A.P. Szczepaniak, Constituent gluons and the static quark potential. Phys. Rev. D **93**(7), 074506 (2016). https://doi.org/10.1103/PhysRevD.93.074506. [arXiv:1505.05104 [hep-lat]]
17. J. Greensite, S. Olejnik, Constituent gluon content of the static quark-antiquark state in Coulomb gauge. Phys. Rev. D **79**, 114501 (2009). [arXiv:0901.0199 [hep-lat]]
18. K. Bardakci, C.B. Thorn, A worldsheet description of large N(c) quantum field theory. Nucl. Phys. B **626**, 287 (2002). [arXiv:hep-th/0110301]
19. C.B. Thorn, A worldsheet description of planar Yang-Mills theory. Nucl. Phys. B **637**, 272 (2002). [Erratum-ibid. B **648**, 457 (2003)]. [arXiv:hep-th/0203167]

The Vacuum Wavefunctional

13

Abstract

The Yang-Mills vacuum wavefunctional as a bridge to magnetic disorder in lower dimensions. Proposals for the Yang-Mills vacuum wavefunctional in 2+1 dimensions. Numerical tests in temporal gauge. The Karabali-Nair "new variables" approach.

The static quark potential arises from vacuum fluctuations of the gauge fields; this is clear from the fact that the potential is extracted from the vacuum expectation value of a Wilson loop. In a Euclidean functional integral, the orientation of a rectangular $R \times T$ loop is obviously irrelevant to the expectation value, and in particular it can be oriented in a plane at a fixed time. In that case

$$W(R, T) = \langle \Psi_0 | \text{Tr}[U(C)] | \Psi_0 \rangle , \tag{13.1}$$

where, in a physical gauge such as temporal $(A_0 = 0)$ or Coulomb gauge, the ground state $\Psi_0[A]$ of the Hamiltonian can be expressed in path-integral form

$$\Psi_0[A(\mathbf{x}, 0)] = \frac{1}{\sqrt{Z}} \int DA(\mathbf{x}, t < 0) \, \delta(F[A]) \det(M) \exp\left[-\int_{-\infty}^{0} dt \, L[A(\mathbf{x}, t)] \right] . \tag{13.2}$$

Here $L[A]$ is the Lagrangian, and $F[A] = 0$ is the gauge-fixing condition with $\det(M)$ the corresponding Faddeev-Popov determinant. In Coulomb gauge an additional restriction of gauge fields to the Gribov region (or some subspace thereof) is understood, as discussed in Chap. 10. A physical gauge is a gauge with a ghost-free Hamiltonian (which requires, among other things, no time-derivatives in the

© Springer Nature Switzerland AG 2020

J. Greensite, *An Introduction to the Confinement Problem*, Lecture Notes in Physics 972, https://doi.org/10.1007/978-3-030-51563-8_13

gauge-fixing condition) and in such gauges Ψ_0 satisfies[1]

$$H\Psi_0[A_k(\mathbf{x})] = E_0\Psi_0[A_k(\mathbf{x})] . \tag{13.3}$$

Then, given (13.1), it seems possible that something could be learned about confinement if we knew the form of the vacuum wavefunctional $\Psi_0[A]$.

13.1   Dimensional Reduction

It was proposed long ago [1] that for gauge-field configurations in which the field strength is weak, and which are "slowly varying" in the sense that gauge-invariant observables vary slowly with respect to the characteristic scale (set by Λ_{QCD}), the vacuum wavefunctional in 3+1 dimensions has the simple form

$$\Psi_0[A] \approx \Psi_0^{eff}[A] = \mathcal{N} \exp\left[-\tfrac{1}{2}\mu \int d^3x \; \text{Tr}[F_{ij}^2]\right] . \tag{13.4}$$

If this form for $\Psi_0^{eff}[A]$ correctly describes the properties of the vacuum at large scales, then the vacuum has the property of *dimensional reduction*, in the sense that the computation of the string tension of a large Wilson loop, in $D = 4$ Euclidean dimensions, reduces to the same calculation in $D = 3$ dimensions with an effective coupling $1/2g_{eff}^2 = \mu$, i.e.

$$\begin{aligned}
W(R, T) &= \langle\text{Tr}[U(C)]\rangle_{D=4} \\
&= \langle\Psi_0|\text{Tr}[U(C)]|\Psi_0\rangle \\
&\approx \int DA(\mathbf{x}) \; \delta(F[A]) \det(M)\text{Tr}[U(C)] \\
&\quad \times \exp\left[-\mu \int d^3x \; \text{Tr}[F_{ij}^2]\right] \\
&= \langle\text{Tr}[U(C)]\rangle_{D=3} .
\end{aligned} \tag{13.5}$$

It was also suggested [2] that the vacuum state of Yang-Mills theory in 2+1 dimensions has, at large scales, the same form (13.4) as in 3+1 dimensions, but in one less spatial dimension. If this is so then dimensional reduction can be applied

---

[1] As already mentioned in Chap. 10, it may still be useful in a physical gauge to introduce Faddeev-Popov ghosts in the path integral, but the essential point is that in a physical gauge these ghosts do not propagate in time.

one more time, and we would find, for large Wilson loops

$$W(R, T) \approx \langle \text{Tr}[U(C)] \rangle_{D=2}$$

$$= \exp[-\sigma \text{Area}(C)] , \qquad (13.6)$$

where the last step follows from the fact that Wilson loops have an area-law falloff in $D = 2$ dimensions.

The proposal (13.4) has been supported in two ways. First of all, it can be shown to be true in strong-coupling lattice gauge theory. In temporal gauge ($U_0 = 1$), the ground state of the lattice Hamiltonian has the form

$$\Psi_0[U] = \mathcal{N} \exp[-R(U)] , \qquad (13.7)$$

and $R(U)$ can be computed systematically in a power series in $1/g^2$ (the procedure is worked out in ref. [3]). At leading order, in $SU(N)$ gauge theory

$$\Psi_0[U] = \mathcal{N} \exp\left[\frac{N}{g^4(N-1)} \sum_P \text{Tr}U(P) + \text{c.c.} \right] , \qquad (13.8)$$

where the sum is over spacelike plaquettes P, and $U(P)$ is the product of links around plaquette P. We see that the exponent has the form of the Wilson action in one lower dimension, and therefore at strong-coupling the dimensional reduction form of the vacuum wavefunctional is seen to be correct.

The second source of support comes from direct measurement, by numerical methods, of the wavefunctional evaluated on a finite set of configurations [4, 5]. This method begins with the identity, in temporal gauge

$$\left| \Psi_0[U_i(\mathbf{x})] \right|^2 = \frac{1}{Z} \int DU_i \, \delta[U_i(\mathbf{x}) - U_i(\mathbf{x}, 0)] e^{-S} . \qquad (13.9)$$

Suppose, in $d + 1$ dimensions, we pick any set of M d-dimensional lattice configurations $\{U_i^m(\mathbf{x}), m = 1, \ldots, M\}$ that may be of interest, and define

$$\tilde{Z} = \sum_{m=1}^{M} \int DU \, \delta[U^m(\mathbf{x}) - U(\mathbf{x}, 0)] e^{-S} . \qquad (13.10)$$

Then $\tilde{Z}$ defines a lattice system in which the configurations at time $t = 0$ are constrained to be members of the given set of M configurations. The statistical probability associated with the k-th configuration in the set is given by

$$\text{Prob}[U^k] = \frac{1}{\tilde{Z}} \int DU(\mathbf{x}, t) \, \delta[U^k(\mathbf{x}) - U(\mathbf{x}, 0)] e^{-S} . \qquad (13.11)$$

In a Monte Carlo simulation of the lattice system described by $\tilde{Z}$,

$$\mathrm{Prob}[U^k] = \lim_{N_{tot} \to \infty} \frac{N_k}{N_{tot}} \,, \tag{13.12}$$

where N_{tot} is the number of times the configuration on the $t = 0$ timeslice is updated by the Metropolis algorithm, and N_k is the number of times the k-th configuration, out of the set $\{U_i^m(\mathbf{x})\}$, $m = 1, 2, \ldots, M$, is generated by the updates. In practice, links at $t \neq 0$ are calculated by some efficient algorithm, such as heat bath for the $SU(2)$ gauge group, while the configuration at $t = 0$ is updated by selecting one of the M configurations at random, and accepting or rejecting the choice via the Metropolis algorithm. With this technique we can also compute the ratio

$$
\begin{aligned}
\frac{|\Psi_0[U^{m_1}]|^2}{|\Psi_0[U^{m_2}]|^2} &= \frac{\int DU \delta[U^{m_1}(\mathbf{x}) - U(\mathbf{x}, 0)]e^{-S}}{\int DU \delta[U^{m_2}(\mathbf{x}) - U(\mathbf{x}, 0)]e^{-S}} \\
&= \frac{\int DU \delta[U^{m_1}(\mathbf{x}) - U(\mathbf{x}, 0)]e^{-S}/\tilde{Z}}{\int DU \delta[U^{m_2}(\mathbf{x}) - U(\mathbf{x}, 0)]e^{-S}/\tilde{Z}} \\
&= \lim_{N_{tot} \to \infty} \frac{N_{m_1}}{N_{m_2}} \,,
\end{aligned}
\tag{13.13}
$$

and in this way it is possible to measure, via a slightly non-standard Monte Carlo simulation, the relative values of the Yang-Mills vacuum wavefunctional in any given set of configurations. The computation was carried out long ago for non-abelian constant configurations (the $U_i(\mathbf{x})$ are independent of $\mathbf{x}$, and $[U_i, U_j] \neq 0$ if $i \neq j$), and also for slowly varying abelian plane wave configurations, with wavelengths comparable to the lattice size. In every case, the results were consistent with the dimensional reduction form

$$\Psi_0[U] = \exp\left[\mu(\beta) \sum_P \mathrm{Tr} U(P) + \text{c.c.} \right] \,, \tag{13.14}$$

where $\mu(\beta)$ in lattice units varies with β in such a way that, when converted to physical units, μ is constant at any β. This behavior was found in both three and four dimensions [4, 5].

The above results also apply to Coulomb gauge, since the Coulomb gauge wavefunctional Ψ_0^{coul} is identical to the temporal gauge wavefunctional Ψ_0

$$\Psi_0^{coul}[A^{tr}] = \Psi_0[A^{tr}] \,, \tag{13.15}$$

when the latter is restricted to transverse gauge fields satisfying the Coulomb gauge condition $\nabla \cdot A^{tr} = 0$ [6]. This means, in particular, that if the temporal gauge vacuum wavefunctional is, at large scales, Gaussian in the field strength F_{ij}, then the Coulomb gauge vacuum wavefunctional is *also* Gaussian in the field strengths.

There is, however, a suggestion by Szczepaniak and Swanson [7, 8] that the vacuum wavefunctional in Coulomb gauge is Gaussian in the gauge fields instead, i.e.

$$\Psi_0^{coul}[A] = \mathcal{N} \exp\left[-\tfrac{1}{2} \int \frac{d^3k}{(2\pi)^3} \, A_i^a(k)\omega(k)A_i^a(-k) \right] , \tag{13.16}$$

with some kernel $\omega(k)$ to be determined by minimization of the vacuum energy. Reinhardt and Feuchter [9, 10] have suggested that this expression should be multiplied by a factor of the inverse square-root of the F-P determinant $M^{-1/2}$. The advantage of these proposals is that a wavefunctional which is Gaussian in the gauge fields, rather than field strengths, lends itself to analytical methods, and an infrared enhancement of the ghost propagator and Coulomb potential has been obtained. The disadvantage is that it is difficult to see how such a wavefunctional could ever lead to an area law for spatial Wilson loops,[2] and the proposal also seems to be at odds with the existing evidence, obtained in temporal gauge, for the dimensional reduction form at large scales.

   However, the dimensional reduction form of the wavefunctional cannot be exact. If it were, then Green's functions in four-dimensional Yang-Mills theory would be identical to those in three-dimensional Yang-Mills theory, which certainly cannot be true at short distances (it would violate perturbation theory, at distance scales where perturbative calculations are valid). Moreover, while reduction to two dimensions seems fine if the Wilson loop is in the fundamental representation, it is clearly wrong for Wilson loops in the adjoint representation, where Wilson loops must ultimately fall off with a perimeter law, due to color screening. The problem is that in two-dimensional Yang-Mills theory the string tension is proportional to the quadratic Casimir of the color representation of the Wilson loop, and therefore we have Casimir scaling rather than N-ality dependence asymptotically. For a more accurate picture of the Yang-Mills vacuum state, it is certainly necessary to go beyond the dimensional reduction form. Most of the recent progress in this area has been in $D = 2 + 1$ dimensions.

13.2   Temporal Gauge Vacuum State in 2+1 Dimensions

The Yang-Mills Hamiltonian in $D = 2 + 1$ dimensions, in temporal gauge, is

$$H = \int d^2x \left\{ -\tfrac{1}{2} \frac{\delta^2}{\delta A_k^a(x)^2} + \tfrac{1}{2} B^a(x)^2 \right\} , \tag{13.17}$$

where $B = F_{12}$, and we will consider, for simplicity, the $SU(2)$ gauge group. A special feature of $A_0 = 0$ gauge is that the Gauss law $D_k^{ab} E_k^b = 0$, which normally follows from stationarity of the action with respect to variations in A_0, must be

---

[2]An attempt to overcome this difficulty by inclusion of monopole fields is in [11].

imposed in the quantum theory as a constraint on physical states, i.e. $D_k^{ab} E_k^b \Psi = 0$. In Schrodinger representation this constraint is

$$\left(\delta^{ac}\partial_k - g\varepsilon^{abc} A_k^b\right)\frac{\delta}{\delta A_k^c}\Psi = 0 ,\qquad(13.18)$$

which is equivalent to requiring the invariance of $\Psi[A]$ under infinitesimal gauge transformations.

The Yang-Mills Schrödinger equation $H\Psi_0 = E_0\Psi_0$ is analytically soluble in two limits: the free-field limit $g \to 0$, in which the vacuum state is found to be

$$\Psi_0[A] = \exp\left[-\frac{1}{2}\int d^2x d^2y\, B^a(x)\left(\frac{\delta^{ab}}{\sqrt{-\nabla^2}}\right)_{xy} B^b(y)\right] ,\qquad(13.19)$$

and also in the zero-mode limit, where one considers only gauge fields which are constant in space.[3] In this limit the Lagrangian is

$$\begin{aligned}L &= \frac{1}{2}\int d^2x\left[\partial_t A_k \cdot \partial_t A_k - g^2(A_1 \times A_2)\cdot(A_1 \times A_2)\right]\\ &= \frac{1}{2}V\left[\partial_t A_k \cdot \partial_t A_k - g^2(A_1 \times A_2)\cdot(A_1 \times A_2)\right]\end{aligned}\qquad(13.20)$$

with corresponding Hamiltonian

$$H = -\frac{1}{2}\frac{1}{V}\frac{\partial^2}{\partial A_k^a \partial A_k^a} + \frac{1}{2}g^2 V(A_1 \times A_2)\cdot(A_1 \times A_2) ,\qquad(13.21)$$

where V is the volume of 2-space, and the cross-product and dot-product are defined with respect to $SU(2)$ color indices, i.e. $A_i = A_i^a \sigma_a/2$, and

$$(A_i \times A_j)^a = \varepsilon_{abc} A_i^b A_j^c .\qquad(13.22)$$

In this case it can be shown that the ground state wavefunction is

$$\Psi_0 = \exp\left[-\frac{1}{2}g V\frac{(A_1 \times A_2)\cdot(A_1 \times A_2)}{\sqrt{|A_1|^2 + |A_2|^2}}\right] .\qquad(13.23)$$

It is not hard to find an expression for $\Psi_0[A]$ which satisfies the Gauss Law constraint (13.18), and also agrees with both soluble limits of the Yang-Mills

---

[3]In quantum gravity this subspace of the set of all configurations is known as "minisuperspace."

Schrodinger equation. It was suggested in [12] that

$$\Psi_0[A] = \exp\left[-\frac{1}{2}\int d^2x d^2y\, B^a(x)\left(\frac{1}{\sqrt{-D^2 - \lambda_0 + m^2}}\right)^{ab}_{xy} B^b(y)\right].$$

$$(13.24)$$

could be a good approximation to the ground state wavefunctional in temporal gauge, where D^2 is the covariant Laplacian, λ_0 is the lowest eigenvalue of $-D^2$, and m^2 is a parameter which vanishes as $g \to 0$.[4] Agreement with the free-field expression in the $g^2 = 0$ limit is obvious. If we instead consider the limit in which the zero modes of the gauge field are much larger than all other modes, then

$$(-D^2)^{ab}_{xy} = g^2\delta^2(x - y)\left[(A_1^2 + A_2^2)\delta^{ab} - A_1^a A_1^b - A_2^a A_2^b\right] \qquad (13.25)$$

and m^2 is negligible. It is then found, after some algebra, that the proposed wavefunctional (13.24) reduces to the zero-mode solution (13.23).

The dimensional reduction form is obtained in the following way: Let

$$(-D^2)^{ab}\phi_n^b(x) = \lambda_n\phi^a(x) \qquad (13.26)$$

denote the eigenvalue equation of the covariant Laplacian, and expand the field strength

$$B^a(x) = \sum_{n=0}^{\infty} b_n\phi_n^a(x). \qquad (13.27)$$

Let us define the "slowly varying" component of the B-field by inserting a cutoff in the mode sum

$$B^{a,\text{slow}}(x) = \sum_{n=0}^{n_{max}} b_n\phi_n^a(x) \qquad (13.28)$$

such that $\lambda_{n_{max}} - \lambda_0 \ll m^2$. Then

$$\int d^2x d^2y\, B^{a,\text{slow}}(x)\left(\frac{1}{\sqrt{-D^2 - \lambda_0 + m^2}}\right)^{ab}_{xy} B^{b,\text{slow}}(y)$$

$$\approx \frac{1}{m}\int d^2x\, B^{a,\text{slow}}(x)B^{a,\text{slow}}(x), \qquad (13.29)$$

---

[4]The subtraction of λ_0 is introduced so that spectrum of $D^2 - \lambda_0 + m^2$ begins at m^2, rather than infinity in the continuum limit. Apart from this subtraction, the proposal is the same as an earlier suggestion by Samuel [13].

and the part of the (squared) vacuum wavefunctional involving only $B^{a,\text{slow}}$ is

$$|\Psi_0|^2 = \exp\left[-\frac{1}{m}\int d^2x \, B^{slow}B^{slow}\right] . \tag{13.30}$$

This has the dimensional reduction form $\Psi_0^2 \sim e^{-S}$, where S is a two-dimensional gauge theory, and the string tension can be computed, in terms of m, analytically. For the $SU(2)$ gauge group this is

$$\sigma = \frac{3}{16}mg^2 , \tag{13.31}$$

or, in terms of the lattice coupling $\beta = 4/g^2$, $\sigma = 3m/(4\beta)$. If we turn this around, and write $m(\beta) = 4\beta\sigma(\beta)/3$, then we have a complete proposal for the Yang-Mills vacuum wavefunctional, in 2+1 dimensions.

To test this proposal, we need to use it to calculate some quantities which can or have been calculated by other means. The string tension is not a test, of course, it is an input. But we could look at other observables, such as the mass gap, computed from the connected equal-times correlator

$$\mathcal{G}(x-y) = \langle (B^a B^a)_x (B^b B^b)_y \rangle - \langle (B^a B^a)_x \rangle^2 \tag{13.32}$$

in the probability distribution

$$P[A] = |\Psi_0[A]|^2 = \exp\left[-\int d^2x d^2y \, B^a(x) K_{xy}^{ab}[A]B^b(y)\right] , \tag{13.33}$$

where

$$K_{xy}^{ab}[A] = \left(\frac{1}{\sqrt{-D^2 - \lambda_0 + m^2}}\right)_{xy}^{ab} . \tag{13.34}$$

Numerically, the computation of $G(x-y)$ looks hopeless. Not only is the kernel $K_{xy}^{ab}[A]$ non-local, it is not even known explicitly for arbitrary $A_i^a(x)$, which would seem to rule out any lattice Monte Carlo approach. But suppose, after eliminating variance along gauge orbits by a further gauge choice (this is allowed in temporal gauge on a given time-slice at $t = 0$), that $K_{xy}^{ab}[A]$ has very little variance among thermalized configurations (i.e. the configurations which are generated by Monte Carlo simulation). In that case, the situation is more promising.

Let us define a probability distribution for gauge fields A which is controlled by a second, independent gauge field A'

$$P\left[A; K[A']\right] = \exp\left[-\int d^2x d^2y \, B^a(x) K_{xy}^{ab}[A']B^b(y)\right] , \tag{13.35}$$

where B is computed from A, not A', and $P[A] = P[A; K[A]]$. Then, assuming the variance of K is small,

$$P[A] \approx P\left[A, \langle K \rangle\right]$$

$$= P\left[A, \int DA'\, K[A']P[A']\right]$$

$$\approx \int DA'\, P\left[A, K[A']\right] P[A'] . \tag{13.36}$$

This expression is amenable to an iterative solution:

$$P^{(1)}[A] = P\left[A; K[0]\right]$$

$$P^{(n+1)}[A] = \int DA'\, P\left[A; K[A']\right] P^{(n)}[A'] . \tag{13.37}$$

The numerical simulation then proceeds in this way: Fix the gauge in the $D = 2$ time-slice to an axial $A_1 = 0$ gauge, and introduce a lattice regularization. Initially, set also $A_2 = 0$. Then

1. Set $A'_2(x) = A_2(x)$.
2. $P[A; K[A']]$ is Gaussian in B. Diagonalize $K^{ab}_{xy}[A']$, and generate a new B-field stochastically.
3. Given B, calculate A_2 in the $A_1 = 0$ gauge, and compute observables.
4. Return to step 1, repeat as many times as necessary.

This procedure will generate configurations with probability distribution $P[A]$ if the kernel K^{ab}_{xy} converges to a fixed matrix (in lattice regularization) in the color-position indices, as the iterations proceed. The lattices which are generated in this way are called *recursion lattices*. Whatever observables are calculated in this way can be compared to a computation of the same (equal-time) observables on lattices generated by the usual Monte Carlo simulation in temporal gauge, with the remaining (time-independent) gauge symmetry used to fix to $A_1 = 0$ gauge on a time-slice at $t = 0$. These lattices at time $t = 0$, are referred to as *MC* lattices, and are drawn from a probability distribution $|\Psi_0^{true}[A]|^2$, where Ψ_0^{true} is true lowest energy eigenstate of the lattice Hamiltonian.[5]

---

[5]More precisely, it is the eigenstate of the lattice transfer matrix $T = \exp[-Ha]$ with the highest eigenvalue.

An important quantity to compute is the correlator

$$G(x - y) = \left\langle (K^{-1})^{ab}_{xy} (K^{-1})^{ba}_{yx} \right\rangle$$
$$K^{-1} = \sqrt{-D^2 - \lambda_0 + m^2} \,, \tag{13.38}$$

which is of interest for two reasons. First of all, the underlying assumption of the numerical simulation is that the variance of K is negligible among recursion lattices; this means that the variance of $K^{-1}K^{-1}$ should also be negligible. Secondly, in the probability distribution governing recursion lattices,

$$\langle B^2(x) B^2(y) \rangle_{conn} \propto G(x - y) \,, \tag{13.39}$$

and the mass gap is extracted from the exponential falloff of $G(R)$ with $R = |x - y|$. Figure 13.1 shows the results for $G(R)$ obtained from ten recursion lattices, and ten MC lattices. The two data sets match almost exactly, but what is equally important is that we find accurate values for $G(R)$ down to magnitudes on the order of 10^{-12}. The only way that this can be possible is that there is virtually no dispersion at all, from one lattice to the next, in the value of $G(x - y)$, and the same must therefore be true for K^{ab}_{xy}. This absence of dispersion verifies the basic assumption underlying the simulation method.

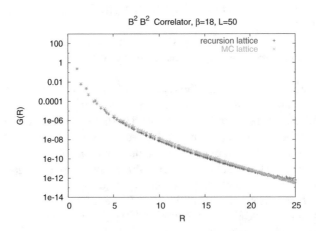

Fig. 13.1 The correlator $G(R)$ computed (1) on two-dimensional lattice configurations generated from the vacuum wavefunctional by the method described in the text; and (2) on constant-time slices of three-dimensional lattice configurations generated by the usual lattice Monte Carlo method. Lattices generated by the first method are denoted "recursion", and by the second as "MC". In each case, the lattice extension is 50 sites at $\beta = 18$. From [12]

In order to extract the mass gap M as a function of β, $G(R)$ is fit to the analytic form ($R = |\mathbf{x} - \mathbf{y}|$)

$$G(R) = \delta^{ab} \delta^{ba} \left[\left(\sqrt{-\nabla^2 + (M/2)^2} \right)_{xy} \right]^2$$

$$= \frac{3}{4\pi^2} (1 + \tfrac{1}{2} M R)^2 \frac{e^{-MR}}{R^6} \, , \qquad (13.40)$$

which in fact fits the recursion lattice result very well, as shown in Fig. 13.2. The value of the mass gaps obtained are plotted in Fig. 13.3, and compared to lattice results for the mass gap obtained by Meyer and Teper [14] in three-dimensional $SU(2)$ lattice gauge theory by standard methods. The agreement is quite good.

Fig. 13.2 Best fit (dashed line) of the recursion lattice data for $G(R)$ by the analytic form given in Eq. (13.40). From [12]

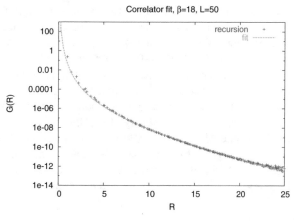

Fig. 13.3 Mass gaps extracted from recursion lattices at various lattice couplings, compared to the 0^+ glueball masses in 2+1 dimensions obtained in ref. [14] (denoted "expt") via standard lattice Monte Carlo methods. Errorbars are smaller than the symbol sizes. From [12]

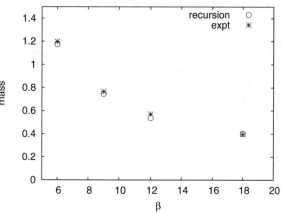

It is also possible to use the vacuum wavefunctional to compute the ghost
propagator in Coulomb gauge

$$G_{gh}(x-y) = \left\langle \left(\frac{1}{-\nabla \cdot D} \right)^{aa}_{xy} \right\rangle , \qquad (13.41)$$

and the color Coulomb potential, which is proportional to

$$\mathscr{V}(x-y) = -\left\langle \left(\frac{1}{\nabla \cdot D}(-\nabla^2)\frac{1}{\nabla \cdot D} \right)^{aa}_{xy} \right\rangle , \qquad (13.42)$$

and to compare these with the values obtained from MC lattices [15]. In either
case, the generated lattice configurations are transformed to Coulomb gauge, and
the observables evaluated in the transformed configurations. The result for the ghost
propagator, obtained from both recursion and MC lattices, is shown in Fig. 13.4.
It is evident that there is almost no difference between the two data sets. The
Coulomb potential is very sensitive to "exceptional" configurations with very small
eigenvalues of the Faddeev-Popov operator $-\nabla \cdot D$; these lead to huge errorbars.
To compare recursion and MC results, we impose cuts on the data, throwing away
these rare configurations. The *same* cuts are applied to both the recursion and MC
data. The aim here is not to get an accurate result for the Coulomb potential, since
a cut on the data can obviously affect the outcome, but rather to see whether the
recursion and MC lattices will yield the same value for the observable, when the
same cut is applied to each set. The result is shown in Fig. 13.5. Obviously, the
results for both the ghost propagator and Coulomb potential obtained from our
proposed wavefunctional closely agree with those obtained by standard methods.

Fig. 13.4 The Coulomb
ghost propagator evaluated on
both recursion and MC
lattices, at lattice coupling
$\beta = 6$ and lattice extension
$L = 24$ in $D = 3$ dimensions.
From [15]

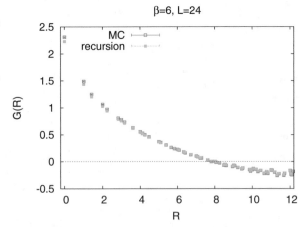

Fig. 13.5 Coulomb potential evaluated from data sets with a cut on lattices with $|V(0)| < -100$, for $\beta = 6$ and lattice extension $L = 24$. Results (with the same cut) are shown for both MC and recursion lattices. From [15]

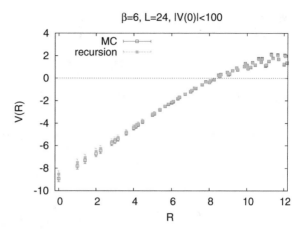

13.3   New Variables

Although the numerical results just presented suggest that the proposal (13.24) is on the right track, it is really only an educated guess for the vacuum wavefunctional in 2+1 dimensions, and the value of the string tension is an input, rather than emerging as an output. A much more ambitious program, initiated by Karabali, Kim, and Nair (KKN) in ref. [16, 17], aims at computing the 2+1 dimensional wavefunctional, and the string tension, systematically. An essential feature of their approach is the transformation to a set of new, gauge-invariant variables.

KKN begin in temporal gauge and combine the remaining gauge fields A_1 and A_2 into a complexified gauge field

$$A \equiv A_z = \tfrac{1}{2}(A_1(x) + i A_2(x)) \ , \quad \overline{A} \equiv A_{\bar{z}} = \tfrac{1}{2}(A_1(x) - i A_2(x)) \ , \qquad (13.43)$$

where $z = x_1 - i x_2$, and $\bar{z} = x_1 + i x_2$ are the usual holomorphic variables in the complex plane. The complexified field A can be related to a matrix field M taking values in the group $SL(2, C)$,

$$A = -(\partial_z M) M^{-1} \ , \quad \overline{A} = M^{\dagger -1} \partial_{\bar{z}} M^{\dagger} \ , \qquad (13.44)$$

which transforms covariantly, $M \to GM$, under a gauge transformation G. Next, define the gauge-invariant variables

$$\mathcal{H} = M^{\dagger} M$$
$$J = \frac{C_A}{\pi} \frac{\partial \mathcal{H}}{\partial z} \mathcal{H}^{-1} \ , \qquad (13.45)$$

where C_A is the quadratic Casimir in the adjoint representation. In terms of these gauge invariant variables, the Hamiltonian becomes

$$H_{KKN} = T + V \, ,   \tag{13.46}$$

where T is derived from the E^2 term in the standard Hamiltonian

$$T = m \left(\int_u J^a(u) \frac{\delta}{\delta J^a(u)} + \int_{u,v} \Omega_{ab}(u,v) \frac{\delta}{\delta J^a(u)} \frac{\delta}{\delta J^b(v)} \right) \, ,   \tag{13.47}$$

with

$$\Omega_{ab}(u,v) = \frac{C_A}{\pi^2} \frac{\delta_{ab}}{(u-v)^2} - i f_{abc} \frac{J^c(v)}{\pi(u-v)}   \tag{13.48}$$

and $(\bar{\partial} \equiv \partial_{\bar{z}})$

$$V = \frac{1}{2g^2} \int B^2(x) = \frac{\pi}{mC_A} \int \bar{\partial} J^a \bar{\partial} J^a   \tag{13.49}$$

and also

$$m = \frac{g^2 C_A}{2\pi} \, .   \tag{13.50}$$

Inner products are evaluated with respect to the integration measure

$$\langle \Psi_1 | \Psi_2 \rangle = \int d\mu(\mathscr{H}) \, e^{2C_A S_{WZW}(\mathscr{H})} \Psi_1^*(\mathscr{H}) \Psi_2(\mathscr{H}) \, ,   \tag{13.51}$$

where $d\mu(\mathscr{H})$ is the standard integration measure for a matrix-valued variable (known as the *Haar measure*), and S_{WZW} is the Wess-Zumino-Witten action [18], whose precise form will not concern us here.

It is important to note that, although the theory is now expressed in terms of gauge-invariant variables $\mathscr{H}$ and J, a new local invariance, *holomorphic invariance*, has appeared in the problem. This is the transformation

$$M(z,\bar{z}) \to M(z,\bar{z}) h^\dagger(\bar{z}) \, , \quad M^\dagger(z,\bar{z}) \to h(z) M^\dagger(z,\bar{z}) \, ,   \tag{13.52}$$

under which

$$H(z,\bar{z}) \to h(z) H(z,\bar{z}) h^\dagger(\bar{z})$$

$$J \to h J h^{-1} + \frac{C_A}{\pi} \partial h h^{-1} \, .   \tag{13.53}$$

It can be seen that J transforms under a holomorphic transformation $h(z)$ much like a gauge field under a gauge transformation. The Hamiltonian H_{KKN} is invariant under these local transformations, and it is crucial that the eigenstates of H_{KKN} are also invariant. In effect, by going to the new variables one trades gauge invariance (in temporal gauge formulation) for holomorphic invariance.

In these new variables, it is possible to carry out a systematic strong-coupling expansion in the continuum formulation. However, since the coupling g^2 has dimensions of mass and simply sets the scale, the question is what parameter controls the convergence of the expansion. KKN suggest that the effective parameter is k/g^2 or (equivalently) k/m, where k is some characteristic momentum in the problem. To leading order, KKN find

$$\Psi_0 = \exp\left[-\frac{\pi}{m^2 C_A} \int d^2x \, \text{Tr}[\bar{\partial} J \bar{\partial} J]\right]$$

$$= \exp\left[-\frac{1}{2mg^2} \int d^2x \, \text{Tr}[B^2]\right] . \tag{13.54}$$

In the usual variables, this ground state has the dimensional reduction form, and spacelike Wilson loops can be computed analytically. These have an area law falloff, and for Wilson loops in the fundamental representation of the $SU(N)$ gauge group, the string tension is

$$\sigma = \frac{g^4}{8\pi}(N^2 - 1) . \tag{13.55}$$

Very remarkably, this value is within a few percent of the value found by Teper and Bringoltz [19] in lattice Monte Carlo simulations of the 2+1 dimensional theory.[6]

On the other hand, we already know that the dimensional reduction form cannot be quite right. It leads to incorrect behavior at high momenta, and string tensions follow the Casimir scaling rule, rather than depending on N-ality. So the question is what is the effect of higher order terms in the perturbation series. KKN manage to sum up all terms in the series, for $\log(\Psi_0)$, which are bilinear in the J's. The result is

$$\Psi_0 = \exp\left[-\frac{2\pi^2}{g^2 C_A^2} \int d^2x d^2y \, \bar{\partial} J^a(x) \left(\frac{1}{\sqrt{-\nabla^2 + m^2} + m}\right)_{xy} \bar{\partial} J^a(y)\right]$$

$$= \exp\left[-\frac{1}{2g^2} \int d^2x d^2y \, B^a(x) \left(\frac{1}{\sqrt{-\nabla^2 + m^2} + m}\right)_{xy} B^a(y)\right] , \tag{13.56}$$

---

[6]Recently some corrections to σ have been calculated [17], and they are small. At present it is not entirely clear *why* the correction is so small, since it involves a sum of rather large (positive and negative) contributing terms, which for some reason nearly cancel.

which has the attractive feature of agreeing with the known result of the free theory at high momenta. The problem with this expression, however, is that in terms of the new variables (top line) it is not holomorphic invariant, and in terms of the old variables in temporal gauge (bottom line) it is not gauge invariant. Because physical states *must* be gauge invariant (this is demanded by the Gauss law constraint), it means that the KKN proposal is incomplete, in the sense that we don't really know what $\Psi_0[A]$ is equal to for an arbitrary gauge field configuration.

There are an infinite number of ways in which (13.56) could be modified such that Ψ_0 becomes gauge-invariant, but reduces to (13.56) in the special case that the gauge field is abelian (i.e. $[A_i, A_j] = 0$). One possibility, suggested by KKN in their original paper, is that the neglected terms in the perturbation series, which involve higher powers of J, will sum up to convert the Laplacian operator ∇^2 to a Laplacian which is covariant under holomorphic transformations (when the wavefunctional is expressed in new variables), or covariant under gauge transformations, when the wavefunctional is converted to old variables. In that case

$$\Psi_0 = \exp\left[-\frac{1}{2g^2} \int d^2x d^2y \; B^a(x) \left(\frac{1}{\sqrt{-D^2 + m^2} + m} \right)_{xy} B^a(y) \right].$$

(13.57)

In this form, the wavefunctional can be latticized and simulated by the technique outlined in the previous section. Unfortunately, when this calculation is performed [12], the resulting string tension differs substantially, by at least 50%, from the known value. The reason is easy to understand: Although the spectrum of $-\nabla^2$ begins at zero, in an infinite volume, the spectrum of the covariant Laplacian $-D^2$ begins at a positive value $\lambda_0 > 0$. Hence the approximation of simply dropping this operator when calculating large-scale Wilson loops, on the grounds that the operator is dominated at long wavelengths by its lowest eigenvalue—namely zero— is unjustified, as shown explicitly by the numerical calculation. This fact implies that the summation of the perturbation series, if done correctly, presumably does *not* lead to the form (13.57). But there are other reasonable gauge-invariant extensions of (13.56) which would avoid this problem, e.g.

$$\Psi_0 = \exp\left[-\frac{1}{2g^2} \int d^2x d^2y \; B^a(x) \left(\frac{1}{\sqrt{-D^2 - \lambda_0 + m^2} + m} \right)_{xy} B^a(y) \right],$$

(13.58)

in which case the KKN wavefunction is rather close in form to (13.24). At present, however, the gauge-invariant extension of (13.56) is not known, so the amplitude of an arbitrary vacuum gauge field is likewise unknown, and the proposal is, in this sense, incomplete.

An ingenious approach to deriving a holomorphic-invariant wavefunctional in the new variables was put forward by Leigh et al. [20]. These authors start from the approximation

$$\Psi_0 = \exp\left[-\frac{\pi}{2C_A m^2}\int d^2x\, d^2y\, \bar{\partial} J^a(x) K_{xy}(L) \bar{\partial} J^a(y)\right],\tag{13.59}$$

where $L = -\Delta/m^2$, and Δ is the holomorphic-covariant Laplacian. They then derive and solve a differential equation for $K(L)$, where L is treated as a number, rather than an operator, and by solving this equation they arrive at

$$K(L) = \frac{1}{\sqrt{L}}\frac{J_2(4\sqrt{L})}{J_1(4\sqrt{L})}.\tag{13.60}$$

If the infrared limit means $L \to 0$, then $K \to 1$, and Ψ_0 has the dimensional reduction form (13.54), leading to a very accurate estimate of the string tension. The correct free-field limit is also obtained. By computing various equal-time correlators of the J field in the vacuum state (13.59), Leigh et al. obtain predictions for the glueball mass spectrum in 2+1 dimensions, which appear to be in remarkable agreement with standard lattice Monte Carlo results.

Although the Leigh et al. approach seems quite successful, two cautionary remarks are in order. First, the differential equation for $K(L)$ is derived by assuming the validity of the operator identity

$$T O_n = (2 + n)m O_n,\tag{13.61}$$

where T is the kinetic term of the KKN Hamiltonian, and

$$O_n = \int \bar{\partial} J \Delta^n \bar{\partial} J.\tag{13.62}$$

This identity has not been proven, however. Secondly, the results for the correlators (and the string tension) are obtained by replacing the holomorphic covariant Laplacian by the ordinary Laplacian, and the effect of this replacement on the values of observables is not entirely clear.

13.4   Numerical Tests of Different Proposals in $D = 2 + 1$ Dimensions

We have already introduced three different proposals for the Yang-Mills vacuum wavefunctional. In $2 + 1$ dimensions these are

$$\Psi_{CG} = \mathcal{N} \exp\left[-\frac{1}{2g^2} \int \frac{d^2k}{(2\pi)^2} A_i^a(-k)\overline{\omega}(k) A_i^a(k) \right].$$

$$\Psi_{GO}[A] = \mathcal{N} \exp\left[-\frac{1}{2g^2} \int d^2x d^2y \, B^a(x) \left(\frac{1}{\sqrt{-D^2 - \lambda_0 + m^2}} \right)_{xy}^{ab} B^b(y) \right]$$

$$\Psi_{hybrid} = \mathcal{N} \exp\left[-\frac{1}{2g^2} \int d^2x d^2y \, B^a(x) \left(\frac{1}{\sqrt{-D^2 - \lambda_0 + m^2} + m} \right)_{xy}^{ab} B^b(y) \right]$$

(13.63)

where in Ψ_{CG} and Ψ_{GO} we have made the rescaling $A_i \rightarrow A_i/g$ so that A_i has units of inverse length, and Ψ_{hybrid} reduces to the KKN proposal when evaluated on abelian configurations. Ψ_{CG} is taken to be the vacuum wavefunctional in Coulomb gauge, and the kernel $\overline{\omega}(k)$ is determined, up to a constant $\overline{\omega}(0) = c_1$, by a set of integral equations derived from minimization of the vacuum energy [7,9].

All of these proposals can be regularized on the lattice, and all then have the form

$$\Psi[U] = \mathcal{N} e^{-\frac{1}{2} R[U]} \tag{13.64}$$

where the $U_i(\mathbf{x})$ are link variables in the space directions. Each proposed vacuum wavefunctional can then be tested by the procedure outlined in Sect. 13.1. As discussed in that section, we choose a set of configurations $\{U_i^{(m)}(\mathbf{x})\}$, $m = 1, 2, \ldots, M$, and carry out a lattice Monte Carlo simulation with the configuration generated at $t = 0$ restricted to members of this set. Let N_m be the number of times that the m-th configuration is chosen in a Monte Carlo update, and let N_{tot} be the total number of updates. The true vacuum wavefunctional of 2+1 Yang-Mills theory will have the property

$$\lim_{N_{tot} \rightarrow \infty} -\log\left(\frac{N_m}{N_{tot}} \right) = R[U^{(m)}] + \text{const.} \tag{13.65}$$

The numerical test is to plot, for some large N_{tot}, the data for $-\log(N_m/N_{tot})$ vs. any proposal for $R[U^{(m)}]$. If the data lies on a straight line with unit slope, that can be regarded as supporting evidence, otherwise the proposal is falsified.

These tests were carried out in ref. [21]. The sets of M configurations were of three types:

- Abelian plane waves with fixed wavelength λ and variable amplitude

$$U_1^{(m)}(n_1, n_2) = \sqrt{1 - (a^{(m)}(n_2))^2}\,\mathbb{1}_2 + ia^{(m)}(n_2)\sigma_3$$

$$U_2^{(m)}(n_1, n_2) = \mathbb{1}_2$$

$$a^{(m)}(n_2) = \frac{1}{L}\sqrt{\alpha + \gamma m}\,\cos\left(\frac{2\pi n_2}{\lambda}\right)\,, \tag{13.66}$$

where $m = 1, 2, \ldots, M$ with L the lattice extension and α, γ some constants. The wavelength can be varied by setting $\lambda = L$ and performing simulations on varying lattice volumes or, alternatively, by setting $\lambda = L/K$, where K is an integer, and carrying out simulations with different values of K on a fixed lattice volume.

- Non-abelian constant configurations, variable amplitude:

$$U_1^{(m)}(n_1, n_2) = \sqrt{1 - (a^{(m)})^2}\,\mathbb{1}_2 + ia^{(m)}\sigma_1$$

$$U_2^{(m)}(n_1, n_2) = \sqrt{1 - (a^{(m)})^2}\,\mathbb{1}_2 + ia^{(m)}\sigma_2$$

$$a^{(m)} = \left[\frac{\alpha + \gamma m}{20L^2}\right]^{1/4}\,. \tag{13.67}$$

- Non-abelian constant configurations, fixed amplitude, variable "non-abelianicity" specified by an angle θ_m

$$U_1^{(m)}(n_1, n_2) = \sqrt{1 - \alpha^2}\,\mathbb{1}_2 + i\alpha\sigma_1$$

$$U_2^{(m)}(n_1, n_2) = \sqrt{1 - \alpha^2}\,\mathbb{1}_2 + i\alpha(\cos(\theta_m)\sigma_1 + \sin(\theta_m)\sigma_2)$$

$$\theta_m = \gamma(m - 1)\pi\,. \tag{13.68}$$

The Monte Carlo calculations were carried out at various lattice couplings β_E for the Euclidean Wilson action, with the lattice scale in physical units, at each β_E set by

$$a = \sqrt{\frac{\sigma_L}{\sigma}} \tag{13.69}$$

where σ_L is the string tension, and σ is assigned (rather arbitrarily, since this is $D = 2 + 1$ dimensions) the value $\sigma = (440\,\text{MeV})^2$. The constants g^2, m in Ψ_{GO} and Ψ_{hybrid} are obtained from a best fit to the N_m/N_{tot} abelian plane wave data, and the string tension, in both cases, can then be obtained from the dimensional reduction values which are

$$\sigma = mg^2 \times \begin{cases} \frac{3}{16} & GO \\ \\ \frac{3}{8} & KKN \end{cases} . \tag{13.70}$$

which come out close ($\sqrt{\sigma} = 460$ and $441\,\text{MeV}$ for the GO and KKN cases, respectively) to the scale setting value of $\sqrt{\sigma} = 440\,\text{MeV}$. All three proposals give a good fit to the abelian plane wave data, but for the CG wavefunctional this requires the choice $\bar{\omega}(0) = 0$, and with that choice the CG wavefunctional is completely insensitive to non-abelian constant configurations.

An example of the data for $-\log(N_m/N_{tot})$ vs R_{GO} of the GO proposal, for non-abelian constant configurations, is shown in Fig. 13.6a; a best linear fit to the data has a slope very close (0.98) to desired value of slope $= 1$. The slopes obtained over a larger range of R, at lattice couplings $\beta_E = 6, 9, 12$, are displayed in Fig. 13.6b. In this figure the R value on the x-axis is the midpoint of the range over which the slope was computed. Very similar results were obtained for the hybrid proposal.

For abelian plane waves, up to the shortest wavelength corresponding to $p^2 = 2.5\,\text{GeV}^2$ that was investigated, the GO and KKN proposals are almost indistinguishable, and both agree very well with the values obtained for the true vacuum wavefunctional (i.e. the data for $-\log(N_n/N_{tot})$), evaluated on these configurations. The Coulomb gauge wavefunctional can also fit the plane wave data with an appropriate choice of parameters, with the choice $\bar{\omega}(0) = 0$. The GO and hybrid wavefunctionals are almost indistinguishable when evaluated on non-abelian constant configurations, and this is probably because they have almost the same dimensional reduction limit. It is found that the GO and hybrid wavefunctionals are in good agreement with the true vacuum wavefunctional for non-abelian constant configurations, as well as for abelian plane waves. The Coulomb gauge

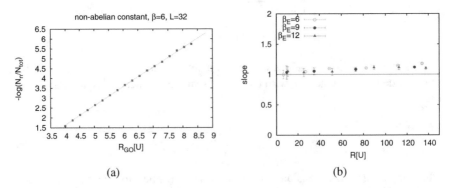

(a)   (b)

Fig. 13.6 (a) Plot of $-\log(N_n/N_{tot})$ vs. $R[U]$ for the Ψ_{GO} wavefunctional, for non-abelian constant configurations of variable amplitude, at $\beta_E = 6$, $L = 32$, $\alpha = 2$, $\gamma = 0.15$. In this case the straight line fit has a slope $= 0.98$. (b) Slopes for the GO wavefunctional vs. R, at $\beta_E = 6, 9, 12$ and $L = 32$, using the values of g^2, m derived from the abelian plane wave fit

wavefunctional, however, which does not have the dimensional reduction property for non-abelian lattices, does not seem compatible with the data for non-abelian constant configurations, particularly the data with variable non-abelianicity.

Further details regarding these numerical tests can be found in ref. [21].

References

1. J. Greensite, Calculation of the Yang-Mills vacuum wavefunctional. Nucl. Phys. B **158**, 469 (1979)
2. M.B. Halpern, Field strength and dual variables formulations of gauge theories. Phys. Rev. D **19**, 517 (1979)
3. J. Greensite, Large scale vacuum structure and new calculational techniques in lattice SU(N) gauge theory. Nucl. Phys. B **166**, 113 (1980)
4. J. Greensite, J. Iwasaki, Monte Carlo study of the Yang-Mills vacuum wave functional in $D = 4$ dimensions. Phys. Lett. B **223**, 207 (1989)
5. H. Arisue, Monte Carlo measurement of the vacuum wave function for non-Abelian gauge theory in D = three-dimensions. Phys. Lett. B **280**, 85 (1992)
6. J. Greensite, S. Olejnik, D. Zwanziger, Coulomb energy, remnant symmetry, and the phases of non-Abelian gauge theories. Phys. Rev. D **69**, 074506 (2004)
7. A.P. Szczepaniak, E.S. Swanson, Coulomb gauge QCD, confinement, and the constituent representation. Phys. Rev. D **65**, 025012 (2002). [arXiv:hep-ph/0107078]
8. A.P. Szczepaniak, Confinement and gluon propagator in Coulomb gauge QCD. Phys. Rev. D **69**, 074031 (2004). [arXiv:hep-ph/0306030].
9. H. Reinhardt, C. Feuchter, On the Yang-Mills wave functional in Coulomb gauge. Phys. Rev. D **71**, 105002 (2005). [arXiv:hep-th/0408237]
10. C. Feuchter, H. Reinhardt, Variational solution of the Yang-Mills Schroedinger equation in Coulomb gauge. Phys. Rev. D **70**, 105021 (2004). [arXiv:hep-th/0408236]
11. A.P. Szczepaniak, H.H. Matevosyan, A model for QCD ground state with magnetic disorder (2010). arXiv:1003.1901 [hep-ph]
12. J. Greensite, S. Olejnik, Dimensional reduction and the Yang-Mills vacuum state in 2+1 dimensions. Phys. Rev. D **77**, 065003 (2008). [arXiv:0707.2860 [hep-lat]]
13. S. Samuel, On the 0++ glueball mass. Phys. Rev. D **55**, 4189 (1997). [arXiv:hep-ph/9604405]
14. H.B. Meyer, M.J. Teper, Glueball Regge trajectories in (2+1)-dimensional gauge theories. Nucl. Phys. B **668**, 111 (2003). [arXiv:hep-lat/0306019]
15. J. Greensite, S. Olejnik, Coulomb confinement from the Yang-Mills vacuum state in 2+1 dimensions (2010). arXiv:1002.1189
16. D. Karabali, C. Kim, V.P. Nair, On the vacuum wave function and string tension of Yang-Mills theories in (2+1) dimensions. Phys. Lett. B **434**, 103 (1998). [arXiv:hep-th/9804132]
17. D. Karabali, V.P. Nair, A. Yelnikov, The Hamiltonian approach to Yang-Mills (2+1): an expansion scheme and corrections to string tension. Nucl. Phys. B **824**, 387 (2010). [arXiv:0906.0783 [hep-th]]
18. E. Witten, Global aspects of current algebra. Nucl. Phys. B **223**, 422 (1983)
19. B. Bringoltz, M. Teper, A precise calculation of the fundamental string tension in SU(N) gauge theories in 2+1 dimensions. Phys. Lett. B **645**, 383 (2007). [arXiv:hep-th/0611286]
20. R.G. Leigh, D. Minic, A. Yelnikov, On the Glueball spectrum of pure Yang-Mills theory in 2+1 dimensions. Phys. Rev. D **76**, 065018 (2007). [arXiv:hep-th/0604060]
21. J. Greensite, H. Matevosyan, S. Olejnik, M. Quandt, H. Reinhardt, A. Szczepaniak, Testing proposals for the Yang-Mills vacuum wavefunctional by measurement of the vacuum. Phys. Rev. D **83**, 114509 (2011). [arXiv:1102.3941 [hep-lat]]

Anti-deSitter Space and Confinement $\qquad$ 14

Abstract

A qualitative introduction to the duality between gauge theories and string theory in anti-deSitter space. Calculation of the string tension in a confining gauge theory from the properties of a black hole horizon in AdS space.

AdS/CFT, also known as the Maldacena conjecture, deserves a book in itself. The initials refer to an equivalence ("duality") between Type II superstrings propagating in a ten-dimensional background space, which is a product of anti-deSitter space (AdS) and some other manifold, with a conformal field theory (CFT). This duality, and its extension (AdS/QCD) to non-conformal, confining gauge theories, is one of the most important results obtained in the modern (i.e. post-1984) era of string theory, and the literature on this topic is vast. Unfortunately, the subject does require a substantial background in superstring theory and conformal field theory, both of which are beyond the scope of this book. Nevertheless, because of the great potential importance of AdS/CFT and AdS/QCD to non-abelian gauge theory and the confinement problem, this chapter aims to provide an inkling of the general idea, together with a sample calculation of Wilson loops via the AdS/CFT approach. The interested reader may wish to consult some of the recent texts on string theory, e.g. [1–3], or specialized review articles [4–6], for a deeper exposition of the subject.

Historically, string theory began as a theory of strings, which are one-dimensional objects propagating in time, but it is now recognized that the theory encompasses other dynamical objects, known as D p-branes (or "D-branes", for short), which are p-dimensional hypersurfaces in space, propagating in time. However, let's begin with strings. The first question is how to represent the motion of a string, and what action should be used, and here it is useful to back up a little further, and consider point particles. The motion of a point particle in $d + 1$ dimensions can be represented by a trajectory $\mathbf{x}(t)$, but this form treats the time coordinate $x^0 = t$ differently from the spatial coordinates x^i, $i = 1, \ldots, d$.

© Springer Nature Switzerland AG 2020

J. Greensite, *An Introduction to the Confinement Problem*, Lecture Notes in Physics 972, https://doi.org/10.1007/978-3-030-51563-8_14

A more natural representation, from the point of view of relativity theory, is $x^\mu(\tau)$, $\mu = 0, 1, \ldots, d$, where τ simply parametrizes the motion. The action of the trajectory $x^\mu(\tau)$ for a free particle of mass m is just the relativistically-invariant proper length L of the trajectory times the mass of the particle:

$$S_{particle} = -mL = -m \int ds$$

$$= -m \int d\tau \sqrt{-\partial_\tau x^\mu \partial_\tau x_\mu} \, . \tag{14.1}$$

The corresponding equations of motion, in flat Minkowski space,

$$p^\nu = \frac{m \partial_\tau x^\nu}{\sqrt{-\partial_\tau x^\mu \partial_\tau x_\mu}} \quad , \quad \partial_\tau p^\nu = 0 \, , \tag{14.2}$$

describe the motion of a free relativistic particle. Of course, there are an infinite number of ways of choosing the parameter τ, since for any *reparametrization* $\tau = \tau(\tau')$, the expressions $x^\mu(\tau)$ and $x^\mu(\tau') \equiv x^\mu(\tau(\tau'))$ describe the same trajectory. It is easy to check that the action $S_{particle}$ is invariant under arbitrary reparametrizations of τ.

A particle is a pointlike object, and traces out a trajectory, or "worldline" as it propagates in time. A string is a one-dimensional object, and therefore it traces out a two-dimensional manifold, or "worldsheet" as it propagates. To specify the location of the worldsheet in the $d + 1$-dimensional background (or "target") space requires two parameters, τ, σ, and it is useful to think of the coordinates $x^\mu(\tau, \sigma)$ of the worldsheet as a field in two-dimensional $\tau - \sigma$ space. The action of the corresponding two-dimensional field theory, i.e. the action of the worldsheet, is defined by analogy to the point particle. Introduce the notation $\sigma^0 = \tau$, $\sigma^1 = \sigma$ and

$$\dot{x}^\mu \equiv \frac{\partial x^\mu}{\partial \tau} \quad , \quad x'^\mu \equiv \frac{\partial x^\mu}{\partial \sigma} \, . \tag{14.3}$$

Since the particle action is proportional to the length of a particle worldline, it is natural that the string action is proportional to the area of a particle worldsheet

$$S = -\mathscr{T} \times \text{Area}$$

$$= -\mathscr{T} \int d^2\sigma \sqrt{-\det[\gamma_{\alpha\beta}]} \, , \tag{14.4}$$

where the constant of proportionality $\mathscr{T}$ must have units of energy per unit length, or, in other words, $\mathscr{T}$ is a string tension. The metric $\gamma_{\alpha\beta}$ (with $\alpha, \beta = 0, 1$) is the two-dimensional induced metric on the world sheet. This induced metric is simple to express in terms of the parametrized worldsheet $x^\mu(\tau, \sigma) = x^\mu(\sigma^\alpha)$. Suppose we have two nearby points on the worldsheet $x^\mu(\sigma^\alpha)$ and $x^\mu(\sigma^\alpha + d\sigma^\alpha)$. Then the squared distance between them is

$$ds^2 = g_{\mu\nu}dx^\mu dx^\nu = g_{\mu\nu}\frac{\partial x^\mu}{\partial \sigma^\alpha}\frac{\partial x^\nu}{\partial \sigma^\beta}d\sigma^\alpha d\sigma^\beta$$

$$= \gamma_{\alpha\beta}[x(\tau,\sigma)]d\sigma^\alpha d\sigma^\beta \,, \tag{14.5}$$

where

$$\gamma_{\alpha\beta} = g_{\mu\nu}\frac{\partial x^\mu}{\partial \sigma^\alpha}\frac{\partial x^\nu}{\partial \sigma^\beta} \,. \tag{14.6}$$

This induced metric gives us the Nambu action

$$S = -\mathscr{T}\int d\tau d\sigma \sqrt{(\dot{x})^2(x')^2 - (\dot{x}^\mu x'_\mu)^2} \,. \tag{14.7}$$

which is the starting point of string theory.

The Nambu action is invariant under the reparametrizations

$$\tau \to \tau(\tau',\sigma') \,, \quad \sigma \to \sigma(\tau',\sigma') \,, \tag{14.8}$$

which allow us to choose, if we wish, some special parametrization. This amounts fixing the reparametrization invariance by a special choice of (or conditions on) the coordinates on the worldsheet, and that choice, in analogy to what is done with ordinary gauge invariance, can be regarded as a choice of gauge. For the calculations in this chapter it is the static gauge, which identifies the x^0 coordinate of a point on the worldsheet with the parameter τ, that will be the most useful.

Strings can be open or closed, and in the case of open strings the equations of motion must be supplemented with boundary conditions. The endpoints of an open string can either be free to move in the d spatial dimensions (Neumann boundary conditions), or each end can be fixed to some definite point in space (Dirichlet boundary conditions). Neumann and Dirichlet boundary conditions may also be mixed, with p coordinates of an endpoint satisfying Neumann conditions, and the rest of the coordinates fixed by Dirichlet boundary conditions. More generally the endpoints of a string may be restricted to move only on a $p + 1$ dimensional hypersurface known as a D p-brane, as illustrated in Fig. 14.1.

If the degrees of freedom of a string are only the coordinates of the string in the $d + 1$-dimensional target space, as is the case for the Nambu action, then all of the quantum excitations of the string (whether open or closed) are bosonic. Because inner products of the vector operators x'^μ and $\dot{x}^\mu$ involve a spacetime metric $g_{\mu\nu}$ which has both positive and negative components, there is a danger of having negative norm states known as "ghosts." This situation is encountered in quantum electrodynamics, where the ghost states are eliminated by imposing a certain operator constraint, related to the Landau gauge condition, on physical states (the Gupta-Bleuler procedure). A similar strategy in string theory only works

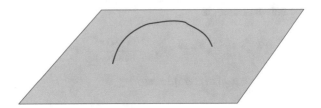

Fig. 14.1 The endpoints of open strings may be restricted to lie on a D p-brane

in $d + 1 = 26$ dimensions. In this critical dimension the spectrum of open bosonic strings (with Neumann boundary conditions) includes a tachyon, with $m^2 < 0$, massless vector particles which can be identified with abelian gauge bosons, and an infinite tower of massive states, lying along Regge trajectories (see Chap. 3) with slope $\alpha' = 1/(2\pi \mathscr{T})$. The spectrum of closed bosonic strings also includes a tachyon, as well as a massless spin-2 state that can be identified as the graviton, and another infinite tower of massive bosonic states. It is not hard to compute scattering amplitudes among these states up to one-loop level. The non-perturbative formulation of bosonic strings is problematic, however, because of the unphysical tachyon states.

Tachyon-free theories can be formulated by introducing fermionic degrees of freedom on the string worldsheet, i.e. there are fermionic fields $\psi(\tau, \sigma)$ as well as bosonic fields $x^\mu(\tau, \sigma)$ on the string worldsheet. Several consistent, tachyon-free string theories (type I, IIA, IIB and heterotic) with fermionic degrees of freedom are known; these are the superstring theories. While these theories were once considered to be entirely independent, they are now understood to be related by various duality transformations, and are believed to be only particular vacua of an underlying theory known as "M-theory." The theory relevant for AdS/CFT is the type IIB theory, which is formulated in terms of interacting closed strings, and is consistent (no negative norm states) in the critical dimension $d + 1 = 10$. If we take the string tension $\mathscr{T} \to \infty$ (or, equivalently, the zero Regge-slope limit $\alpha' \to 0$) in the type IIB theory, then the mass of the massive string excitations goes to infinity, and only the zero mass particles remain. These include the graviton, and some other massless fermions and bosons which fill out the supermultiplet of type IIB supergravity in ten dimensions.

D-branes enter the picture in an interesting and unexpected way. They are not merely allowable boundary conditions restricting the motion of open string endpoints, but turn out to be physical objects in their own right, with their own dynamics. There are several reasons for this. In the first place, string interactions include gravitational interactions, and having an object at a fixed point in space, insensitive to the passage of gravitational waves or any other gravitational phenomena, is simply inconsistent with general relativity. D-branes must, therefore, be dynamical. Secondly, the zero-slope limit of superstring theories, which are various types of supergravity theories, contain p-branes as classical, solitonic solutions of the equations of motion (much as the Georgi-Glashow theory contains magnetic monopoles), and these possess a certain type of charge which can be identified with the charge of D p-branes in string theory. Finally, it was discovered in the 1990s that all the known superstring theories with spatial directions compactified on a circle ("toroidal compactification") are related by various duality transformations, and D p-branes arise inevitably as a consequence of some of these transformations.

D-branes also have an interesting effective action at low energies [7]. Among the quantum states of open strings whose endpoints lie on D-branes are massless vector particles, i.e. photon excitations. Therefore, electromagnetic fields can live on D-branes. Let us identify the first $p + 1$ coordinates $\sigma^a = x^a$, $a = 0, 1, \ldots, p$ as coordinates along the brane worldvolume, and the remaining coordinates x^m, $m > p$ are in directions transverse to the brane. This is the static gauge for the brane. Then in a bosonic theory, the low-energy D-brane dynamics is described by the Dirac-Born-Infeld action

$$S_{DBI} = -\mathcal{T}_{brane} \int d^{p+1}\sigma \sqrt{-\det(\eta_{ab} + \partial_a x^m \partial_b x_m + 2\pi\alpha' F_{ab})}\,, \tag{14.9}$$

where η_{ab} and F_{ab} are the Minkowski metric and electromagnetic field strength, respectively, on the brane, so the argument of the determinant is a $(p+1) \times (p+1)$ matrix. Expanding the square root determinant in powers of $\partial_a x^m$ and F_{ab}, one finds to leading order

$$S_{DBI} \approx -\tfrac{1}{4}\mathcal{T}_{brane}(2\pi\alpha')^2 \int d^{p+1}\sigma \left(F_{ab}F^{ab} + \frac{2}{(2\pi\alpha')^2}\partial_a x^m \partial^a x_m \right)$$
$$- \mathcal{T}_{brane} V_{p+1} + O(F^4)\,, \tag{14.10}$$

where V_{p+1} is the worldvolume of the p-brane. To this lowest order, the D-brane action is simply Maxwell electrodynamics plus some real scalar fields in $p + 1$ dimensions.

14.1   The Maldacena Conjecture

Consider some set of N D 3-branes whose locations exactly coincide. Each endpoint of an open string can lie on any one of the N D p-branes, which makes N^2 massless vector particles in all. In bosonic string theory, the low-energy dynamics of the brane turns out to be $U(N)$ gauge theory,[1] and in the case of superstrings it is $\mathcal{N} = 4$ super Yang-Mills theory in 3+1 dimensions. The latter is a gauge theory with N colors and matter content such that the beta function is zero, which means that the theory is conformally invariant, and non-confining. Conformal invariance is a symmetry which implies, among other things, that there are no dimensionful

---

[1] Which is really $U(1) \times SU(N)$ gauge theory, since one gauge boson decouples from the rest.

constants in the theory, such as the QCD Λ-parameter, or the gauge coupling constant itself in $d + 1 \neq 4$ dimensions.

The relevant constants, in this setup of coinciding D 3-branes, are the following:

- The string tension (energy/unit length) of strings in type IIB string theory $\mathscr{T} = 1/(2\pi\alpha')$. This sets the scale for the masses of the massive string excitations.
- The string coupling constant g_s of type IIB string theory. This constant controls the strength with which strings interact with each other.
- The coupling constant for the $\mathscr{N} = 4$ super Yang-Mills theory, $g^2 = 4\pi g_s$;
- The number of N of coinciding D 3-branes;
- The $d = 10$ dimensional Newton constant $16\pi G = (2\pi)^7 g_s^2 \alpha'^4$.

The AdS/CFT connection begins with an observation about the low-energy excitations of type IIB string theory, in $d = 10$ dimensions, containing a stack of coinciding D 3-branes. The observation is that there are two quite different descriptions of these low-energy excitations in two different limits, depending on the strength of the string coupling constant g_s. Let's begin with the weak-coupling limit. D-branes are massive, and since the low-energy limit of type IIB string theory is type IIB supergravity, these D-branes should (like any massive source), curve spacetime. However, there are N of these branes, and Newton's constant is proportional to g_s, so in the limit that $g_s N \ll 1$ the effect of the branes on the spacetime background should be negligible. Also, in the limit that g_s is small, strings hardly interact with each other. So at low energies the excitations in the theory are of two kinds (see Fig. 14.2):

1. the massless excitations of the strings tied to the D-branes, and their interactions, which is described by $\mathscr{N} = 4$ super Yang-Mills theory;
2. very weakly interacting gravitons, and their superpartners (which make up the spectrum of type IIB supergravity), propagating in flat $D = 10$ dimensional Minkowski space.

Now consider the opposite limit, with $g_s N \gg 1$, but still $g_s \ll 1$. Here the effect of the N D-brane sources on the background geometry cannot be ignored, and in this limit that effect can be treated semiclassically, just by solving the field equations of supergravity (a generalization of Einstein's equations) with the static D-brane source. The solution for the $d = 10$ dimensional spacetime metric turns out to be

$$ds^2 = f^{-1/2}\left(-dt^2 + \sum_{i=1}^{3} dx_i^2\right) + f^{1/2}(dr^2 + r^2 d\Omega_5^2)$$

$$f = 1 + \left(\frac{R}{r}\right)^4$$

$$R^2 = \alpha'\sqrt{4\pi g_s N} , \tag{14.11}$$

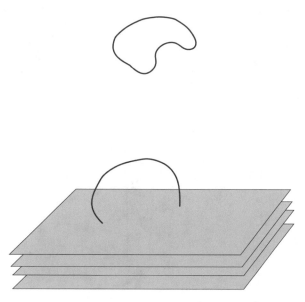

Fig. 14.2 First description, at $g_s N \ll 1$, of the low energy excitations in a ten-dimensional space containing a stack of N parallel D 3-branes. These are $\mathcal{N} = 4$ super Yang-Mills theory (from strings on the branes) and non-interacting massless particles (a supergravity multiplet) in the "bulk" ten-dimensional Minkowski space

whose salient features are represented in Fig. 14.3. At $r \to \infty$, the geometry is simply flat Minkowski space. But as $r \to 0$ (where the stack of D-branes are located), the geometry develops a long "throat", and the circles around the throat represent a five-dimensional sphere S^5, whose radius R, far down the throat, is almost constant. The type IIB string excitations of this system, which have very low energy with respect to an observer in the flat Minkowski region (i.e. at very large r), are of two sorts. First, there are the massless excitations of the string in the large r Minkowski region. On dimensional grounds, if the strings have energy E, they would interact with strength GE^8, where G is the ten-dimensional Newton constant. Therefore, the low-energy closed strings are essentially non-interacting, and moreover, because their wavelengths are much greater than the radius of the throat, they are insensitive to the excitations deep inside the throat region. The other type of low energy excitations in this geometry, as seen by the observer in the asymptotic region, are the string excitations deep within the throat region. These excitations are so strongly red-shifted, with respect to the outside observer in the asymptotic, Minkowski-space region, that they appear to have an arbitrarily low energy, as seen by that observer, providing the string is sufficiently near the horizon at $r = 0$. Therefore, in this opposite $g_s N \gg 1$ limit, the low energy excitations of the theory again fall into two categories:

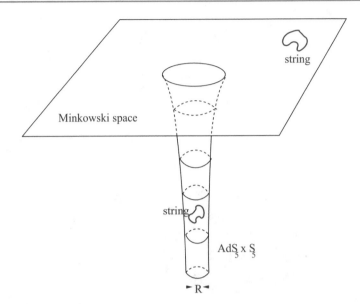

Fig. 14.3 Second description, at $g_s N \gg 1$, of the low energy excitations in a ten-dimensional space with a p=3 brane source. These are type IIB superstrings close to the $r = 0$ horizon, and non-interacting massless particles (a supergravity multiplet) in the asymptotic ten-dimensional Minkowski space

1. Type IIB superstrings close to the $r = 0$ horizon;
2. very weakly interacting gravitons, and their superpartners, propagating in the asymptotically flat Minkowski space.

Both the $g_s N \ll 1$ and $g_s N \gg 1$ limits include weakly interacting gravitons in the low energy limit. However, the other class of low energy excitations is $\mathcal{N} = 4$ super Yang-Mills theory, at $g_s N \ll 1$, and type IIB superstrings in the near-horizon (or "throat") region of the metric in (14.11), at $g_s N \gg 1$. Maldacena [8] made the important conjecture that $\mathcal{N} = 4$ super Yang-Mills theory is equivalent ("dual") to type IIB superstrings propagating in the throat region, so that the latter theory, at $g_s N \gg 1$, is somehow describing $\mathcal{N} = 4$ super Yang-Mills theory at strong couplings. The metric in the throat region, far from the asymptotic Minkowski space region, is obtained by simply neglecting the constant 1 in the definition of $f(r)$, i.e.

$$ ds^2 = \frac{r^2}{R^2} \left(-dt^2 + \sum_{i=1}^{3} dx_i^2 \right) + \frac{R^2}{r^2} (dr^2 + r^2 d\Omega_5^2) , \qquad (14.12) $$

which is the product of anti-deSitter space in five spacetime dimensions AdS_5 and the five-sphere S_5. Thus the conjecture: $\mathcal{N} = 4$ *super Yang-Mills theory in 3+1 dimensions (which is a conformal field theory) is dual to type IIB string theory in an $AdS_5 \times S_5$ spacetime background.* This is actually just one particular example of the AdS/CFT correspondence; other dualities follow from other arrangements of D p-branes.

To make this conjecture precise, it is still necessary to specify what is meant by the correspondence; i.e. what does "duality" mean, exactly? The basic idea, first of all, is that to every string field $\Phi(x, r)$ (one of the string excitation modes such as the graviton, dilaton, etc.) there corresponds a local, gauge-invariant observable $\mathcal{O}(x)$ in the gauge theory. Let $\phi(x)$ be a source term for the operator $\mathcal{O}(x)$. The AdS/CFT correspondence then says that

$$\left\langle \int d^4x \, \phi(x)\mathcal{O}(x) \right\rangle_{CFT} = Z_{string}\left[\Phi(x, r = \infty) = \phi(x) \right], \qquad (14.13)$$

where the left hand side is the expectation value in the gauge theory, while the right hand side is the partition function for the string theory, with the string field $\Phi(x, r)$ held fixed and equal to $\phi(x)$ on the boundary. There is a kind of dictionary relating operators $\mathcal{O}(x)$ in $\mathcal{N} = 4$ super Yang-Mills theory to string fields $\Phi(x, r)$ on $AdS_5 \times S_5$, which we cannot go into here.

In order to use the AdS/CFT correspondence to calculate observables in the gauge theory, it is necessary to be able to solve the string theory in the $AdS_5 \times S_5$ background spacetime. This is a difficult thing to do, in general. In order to make the string theory tractable, we require that the string coupling constant is small, i.e. $g_s \to 0$, and that the radius of curvature R of $AdS_5 \times S_5$ spacetime is very large compared to the length scale characteristic of strings, which is set by the Regge slope α'. In these two limits, the string theory is well approximated by its supergravity limit, which can be evaluated semiclassically, i.e.

$$Z_{string} \sim e^{-S_{sugra}}, \qquad (14.14)$$

where S_{sugra} is the (on-shell) supergravity action. But then, since

$$\frac{R^4}{\alpha'^2} \sim g_s N \gg 1, \qquad (14.15)$$

it means that this semiclassical approximation also requires taking the large-N limit. However, when one analyzes the leading (planar) diagrams in any $SU(N)$ gauge theory, it turns out that the effective coupling is not g^2, but actually $\lambda = g^2 N$, as discussed in Chap. 12. Because $g^2 \sim g_s$, the condition (14.15) means that a calculation on the string theory side in the supergravity limit provides information about the gauge theory in the strong 't Hooft coupling $\lambda \gg 1$ regime, which is not easily accessible by any other method.

14.2   Wilson Loops in AdS Space

Consider a loop C at the boundary of AdS_5 space, i.e. at $r = \infty$, and the worldsheet of a superstring which has C as its boundary. Let $S_{sheet}(C)$ be the action of this worldsheet at the saddlepoint (i.e. the minimal area string configuration). It was argued by Maldacena [9] that in the limit of large N and large 't Hooft coupling λ, where the semiclassical, supergravity approximation to string theory is valid, the gauge/string correspondence for Wilson loops would be

$$\left\langle \text{Tr} P \exp[i \oint_C dx^\mu A_\mu] \right\rangle \approx \exp[-S_{sheet}(C)] . \tag{14.16}$$

Choosing rectangular $L \times T$ loops, with $T \gg L$, the static quark potential can be extracted from Wilson loops in the usual way. On the gauge theory side, one can calculate the left hand side of (14.16) perturbatively at small 't Hooft coupling $\lambda \ll 1$, and determine the potential $V(L) \sim -\lambda/L$. On the string theory side, the right hand side of (14.16) can be computed at large $\lambda \gg 1$, with the result that $V(L) \sim -\sqrt{\lambda}/L$. In either limit the potential is non-confining, as expected in a conformal theory such as $\mathcal{N} = 4$ super Yang-Mills. Since there is no dimensionful scale in a conformal theory, there is nothing to set the scale of the string tension; it can only be zero. Therefore, on dimensional grounds, the only possibility is that $V(L) \sim -1/L$. In order to use the gauge/string correspondence to investigate confining theories, conformal invariance has to be broken in some way, and a dimensionful scale introduced.

Conformal invariance can be broken in a number of ways, and one possibility, suggested by Witten [10], is to introduce a finite temperature. On the gauge theory side, a finite temperature can be introduced by going to a Euclidean metric signature and toroidally compactifying the time variable, so that the periodic time direction has some finite extension β. Then it is necessary to impose periodic boundary conditions on the bosons, and antiperiodic boundary conditions on the fermions. While this leaves some zero-energy bosonic modes, there are no zero energy fermion modes, and supersymmetry is completely broken. At scales much larger than the time compactification scale the theory is effectively three, rather than four dimensional, and the scalars acquire a large mass from one-loop quantum effects. Effectively, the $\mathcal{N} = 4$ super Yang-Mills theory will appear, at large scales, like a pure, non-supersymmetric Yang-Mills theory in three Euclidean dimensions, and this theory is, of course, confining.

On the string theory side it is also possible to go to a Euclidean signature and compactify the time direction. Then, as shown by Hawking and Page [11], there exist black hole solutions in AdS spacetime. There is a temperature on the string side as well: it is the Hawking temperature which is associated with any black hole horizon. The relevant metric is

$$ds^2 = \alpha' \left\{ \frac{U^2}{\tilde{R}^2}[h(U)dt^2 + dx_i^2] + \frac{\tilde{R}^2}{U^2} h^{-1}(U)dU^2 + \tilde{R}^2 d\Omega_5^2 \right\} , \tag{14.17}$$

where we have rescaled

$$U = \frac{r}{\alpha'} \ , \quad \tilde{R} = \frac{R}{\sqrt{\alpha'}} \tag{14.18}$$

and

$$h(U) = 1 - \frac{U_T^4}{U^4} \ , \quad \tilde{R}^2 = \sqrt{4\pi g_s N} \ . \tag{14.19}$$

The black-hole horizon is at $U = U_T$, and the corresponding Hawking temperature is $U_T/(\pi \tilde{R}^2)$. The conjecture is that type IIB string theory in this background is dual to $\mathcal{N} = 4$ super Yang-Mills theory at high temperature, which at distance scales much larger than β should be equivalent to $D = 3$ dimensional (non-supersymmetric) Yang-Mills theory at zero temperature. Then the computation of a large Wilson loop in $D = 3$ Yang-Mills theory, at large 't Hooft coupling, can be solved by finding the minimal area of the string worldsheet bounded by the curve C at the boundary $U = \infty$ of the AdS space.

The calculation goes as follows [12]: Consider rectangular $L \times T$ contours C, which lie in the $x - x'$ plane (e.g. $x = x^1, x' = x^2$) at some fixed point in S_5, and at the boundary $U = \infty$ of the AdS space. It is convenient to adopt a static gauge in which worldsheet coordinates σ, τ are identified with the spacetime coordinates $\sigma = x$, $\tau = x'$. It is assumed that $T \gg L$, and the string is static in the sense that the transverse position $U(\sigma, \tau)$ of the worldsheet depends only on x. Using the ten-dimensional metric $g_{\mu\nu}$ from the line element (14.17) to contract spacetime indices in the Nambu action, and integrating over $\tau = x'$

$$S = \frac{T}{2\pi} \int_{-L/2}^{L/2} \frac{dx}{\sqrt{h(U)}} \sqrt{(\partial_x U)^2 + (U^4 - U_T^4)/\tilde{R}^4} \ . \tag{14.20}$$

Since the Lagrangian in (14.20) does not depend on x explicitly, there is a constant of motion (if we think of x as "time," the constant of motion is analogous to energy)

$$\frac{U^4 - U_T^4}{\sqrt{h(U)}\sqrt{(\partial_x U)^2 + (U^4 - U_T^4)/\tilde{R}^4}} = \text{constant} \ . \tag{14.21}$$

If we evaluate this constant at $x = 0$, where by $U(x) = U(-x)$ symmetry we have $\partial_x U = 0$, and let $U_0 = U(x = 0)$ denote the minimal value of $U(x)$ in the interval $x \in [-\frac{1}{2}L, \frac{1}{2}L]$, then

$$\frac{U^4 - U_T^4}{\sqrt{h(U)}\sqrt{(\partial_x U)^2 + (U^4 - U_T^4)/\tilde{R}^4}} = \frac{\tilde{R}^2}{\sqrt{h(U_0)}}\sqrt{U_0^4 - U_T^4} \ . \tag{14.22}$$

Solving for $\partial_x U$, and writing

$$y = U/U_0, \quad \varepsilon = h(U_0) = 1 - \frac{U_T^4}{U_0^4}, \tag{14.23}$$

one finds

$$x = \frac{\widetilde{R}^2}{U_0} \int_1^{U/U_0} \frac{dy}{\sqrt{(y^4 - 1)(y^4 - 1 + \varepsilon)}}, \tag{14.24}$$

which determines $U(x)$ implicitly, given U_0. But $U_0 = U(0)$, while the ends of the string are at $x = \pm L/2$, where $U = \infty$, so that

$$L = 2\frac{\widetilde{R}^2}{U_0} \int_1^{\infty} \frac{dy}{\sqrt{(y^4 - 1)(y^4 - 1 + \varepsilon)}} \tag{14.25}$$

gives U_0 as a function of quark separation L.

The energy of the static worldsheet $E = S/T$ is calculated from (14.20) by using the relationships (14.22), and changing integration variable from x to y via (14.24)

$$E(L) = \frac{U_0}{\pi} \int_1^{\infty} dy \frac{y^4}{\sqrt{(y^4 - 1)(y^4 - 1 + \varepsilon)}}. \tag{14.26}$$

It is clear that this energy is infinite, but we should recall that the expectation value of a Wilson loop in continuum gauge theory vanishes (and hence the static quark potential is infinite), due to the infinite self-energy of a static charge. In order to extract the L-dependence of the potential, it is necessary to discard the infinite, and L-independent, self-energy. Following Maldacena [8] we regularize by integrating y only up to $y_{max} = U_{max}/U_0$, and then subtract

$$E_0 = \frac{U_T}{\pi} \int_1^{U_{max}/U_T} dy\, 1$$

$$= \frac{U_{max} - U_T}{\pi}, \tag{14.27}$$

so that in the $U_{max} \to \infty$ limit

$$E_{int} = E(L) - E_0$$

$$= \frac{U_0}{\pi} \int_1^{\infty} dy \left(\frac{y^4}{\sqrt{(y^4 - 1)(y^4 - 1 + \varepsilon)}} - 1 \right) - \frac{U_0 - U_T}{\pi}. \tag{14.28}$$

At any finite L there is a corresponding finite $U_0 > U_T$, with $U_0 \to U_T$ in the $L \to \infty$ limit, and with energies given by the expression in (14.26). Some typical

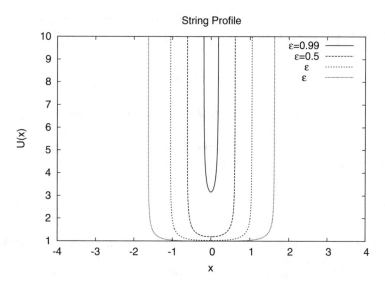

Fig. 14.4 String contours $U(x)$ for various $\varepsilon = 1 - (U_T^4/U_0^4)$ and $\widetilde{R} = 1$, with units chosen such that $U_T = 1$. The asymptotes of each curve lie at $x = \pm L/2$. Note the approach to the horizon (here at $U = 1$), as $\varepsilon \to 0$

solutions for $U(x)$ at various U_0/U_T, are shown in Fig. 14.4. Note that as $\varepsilon \to 0$ and $L \to \infty$ the string worldsheet flattens out along the horizon $U(x) = U_T$, in the finite interval $|x| < L/2$.

For small interquark separation L, the minimal value U_0 is much larger than U_T, and $\varepsilon \approx 1$. In the $\varepsilon = 1$ limit all of the integrals can be carried out analytically, and $L \propto 1/U_0$, while $E_{int}(L) \propto -U_0$. From this, it is clear that $E_{int}(L) \propto -1/L$. The precise result is

$$E(L) = -\frac{1}{L} \frac{4\pi^2 (2g^2 N)^{1/2}}{\Gamma(\frac{1}{4})^4} \; . \tag{14.29}$$

For large interquark separation, we have $U_0 \to U_T$ (and $\varepsilon \to 0$) as $L \to \infty$ for spacelike loops. Both L and $E(L)$ diverge, for spacelike loops, as $-\log(\varepsilon)$. The main contribution to the integrals in (14.25) and (14.26) comes from the integration region near $y = 1$, where the integrands are identical. Extracting a constant of proportionality by inspection, we find

$$E_{int}(L) = \frac{U_T^2}{2\pi \widetilde{R}^2} L \; . \tag{14.30}$$

The result is easy to interpret: the coefficient of L is simply the action density for a worldsheet lying on the horizon, in the $x - x'$ plane, at $U = U_T$, i.e.

$$E_{int}(L) = \frac{1}{T} \lim_{U \to U_T} S[U, \partial_x U = 0] = \frac{U_T^2}{2\pi \widetilde{R}^2} L . \tag{14.31}$$

This makes sense because, as is clear from Fig. 14.4, the minimal area string worldsheet falls from $U = \infty$ to $U = U_T$, and flattens out along the horizon as $L \to \infty$ ($\varepsilon \to 0$). As a result, the string tension for spacelike loops is essentially the action density of a worldsheet on the horizon.

14.3 AdS/QCD

We have arrived at a truly remarkable conclusion: confinement on the gauge theory side can be derived from the corresponding AdS black hole metric on the string theory side. The main requirement was to work with an AdS geometry, on the string side, dual to a non-conformal gauge theory, and the high temperature approach suggested by Witten is only one possibility. Other setups of branes in the $D = 10$ dimensional space, suggested by Polchinski and Strassler [13], Klebanov and Strassler [14], Maldacena and Nunez [15], DiVecchia et al. [16], and others, are also dual to confining gauge theories. The proposal by Sakai and Sugimoto [17], which is dual to a gauge theory including quarks, is perhaps the closest to realizing QCD itself. This general approach, which begins with a particular setup of D-branes and then works out the ten-dimensional geometry and the corresponding gauge theory, is sometimes referred to as the "top-down" approach to AdS/QCD. The "CFT" initials are no longer really appropriate, since the dual gauge theories are non-conformal and confining.

There is an alternative philosophy, known as the "bottom-up" approach to AdS/QCD. Here the idea is just to cook up a field theory, in some truncated version of AdS space, from which one could "solve" QCD, using the field-operator correspondence. The fields that are taken to propagate in the AdS space are chosen based on the known properties of QCD, and the AdS geometry is modified (by hand) in such a way, for example by truncation of the space near $r = 0$, that the string worldsheet calculation will lead to an area law for Wilson loops. This is obviously a phenomenological, rather than fundamental approach, but it has had some impressive numerical successes. For an assessment of the promise and the drawbacks of this approach, cf. [18] and [19].

14.3.1 AdS Embedding of Conformal Quantum Mechanics

An interesting model of QCD hadrons, with some connection to AdS holography, has been developed by de Teramond et al. [20]. Here we can only sketch the general idea. In order to calculate the meson mass spectrum, these authors begin with conformal quantum mechanics, with a time evolution operator

$$G = \frac{1}{2} u \left(-\frac{d^2}{dx^2} + \frac{g}{x^2} \right) + \frac{i}{4} v \left(x \frac{d}{dx} + \frac{d}{dx} x \right) + \frac{1}{2} w x^2 \tag{14.32}$$

where u, v, w are undetermined constants, and this is matched to a light front bound state equation

$$\left(-\frac{d^2}{d\zeta^2} - \frac{1 - 4L^2}{4\zeta^2} + U(\zeta, J) \right) \phi(\zeta) = M^2 \phi(\zeta), \tag{14.33}$$

by identifying $x = \zeta$ and setting

$$u = 2 \ , \quad v = 0 \ , \quad g = L^2 - \frac{1}{4} \ , \quad w = 2\lambda_M^2 \ , \quad U \sim \lambda_M^2 \zeta^2 \tag{14.34}$$

The variable ζ represents the separation of quarks at equal light front time. Confinement in this setup is built in from the start, with a quadratic potential in the ζ coordinate.

The next step is to embed this quantum mechanical model, which combines elements of conformal quantum mechanics and light front kinematics, in a holographic setting. One begins from the effective action for a spin-J field in AdS_{d+1} space represented by a totally symmetric rank-J tensor field $\Phi_{N_1 \ldots N_J}$, where M, N are the indices of the $d + 1$ higher dimensional AdS space with coordinates $x^M = (x^\mu, z)$

$$S_{eff} = \int d^d x \, dz \sqrt{g} \, e^{\varphi(z)} \, g^{N_1 N_1'} \cdots g^{N_J N_J'}$$
$$\left(g^{MM'} D_M \Phi^*_{N_1 \ldots N_J} D_{M'} \Phi_{N_1' \ldots N_J'} - \mu_{eff}^2(z) \, \Phi^*_{N_1 \ldots N_J} \, \Phi_{N_1' \ldots N_J'} \right), \tag{14.35}$$

where x^μ, z are the Minkowski and holographic coordinates, respectively, $\sqrt{g} = (R/z)^{d+1}$ and D_M is the covariant derivative which includes the affine connection (R is the AdS radius). The dilaton $\varphi(z)$ breaks the maximal symmetry of AdS. The association to the light front wavefunction is made through the identification

$$\Phi_{\nu_1 \cdots \nu_J}(x, z) = e^{iP \cdot x} \, \varepsilon_{\nu_1 \cdots \nu_J}(P) \, \Phi_J(z), \tag{14.36}$$

with invariant hadron mass $P_\mu P^\mu \equiv \eta^{\mu\nu} P_\mu P_\nu = M^2$. Variation of the action leads to the wave equation

$$\left[-\frac{z^{d-1-2J}}{e^{\varphi(z)}} \partial_z \left(\frac{e^{\varphi(z)}}{z^{d-1-2J}} \partial_z \right) + \frac{(\mu R)^2}{z^2} \right] \Phi_J = M^2 \Phi_J, \tag{14.37}$$

where $(\mu R)^2 = (\mu_{eff}(z) R)^2 - J z \varphi'(z) + J(d - J + 1)$ is a constant determined by kinematical conditions in the light front. With the dilaton field as $\varphi = \lambda_M z^2$, carrying out some further rescalings

$$\Phi_J(z) = (R/z)^{J-(d-1)/2} e^{-\varphi(z)/2} \phi_J(z). \tag{14.38}$$

and identifying the holographic coordinate z with the light front coordinate ζ, Brodsky et al. arrive at a light front potential

$$U(\zeta, J) = \lambda_M^2 \zeta^2 + 2\lambda_M(J-1). \tag{14.39}$$

These authors then solve (14.33) with the above potential to determine the mass spectrum

$$M_{n,J,L}^2 = 4\lambda_M \left(n + \frac{J+L}{2} \right), \tag{14.40}$$

which features a zero pion mass in the chiral limit, and linear Regge trajectories with the same slope in the quantum numbers n and L.

The baryon mass spectrum is obtained by a similar procedure, this time starting from superconformal quantum mechanics, and embedding the theory in an effective action in AdS$_5$ for fields of half-integer spin. The baryon resonances again fall on linear Regge trajectories in good agreement with experimental data.

It should be understood that this conformal light front approach is only a model of the hadrons in QCD, not QCD itself, and in fact this is really quantum mechanics rather than quantum field theory. But with that reservation, it is also fair to say that this is still an extraordinarily successful model, with a mass spectrum, for both mesons and baryons, that is in impressive agreement with experiment.

14.3.2 Discussion

Returning to the top-down models: do these prove confinement in non-abelian gauge theories? The answer is no, or at least, not yet. First of all, the Maldacena conjecture itself has not yet been proved, although there is a great deal of evidence in its favor (cf. the cited texts and review articles). Apart from that, the main limitation seems to be that a controlled calculation of the Yang-Mills string tension on the string theory side requires a strong 't Hooft coupling $\lambda = g^2 N \gg 1$. In this sense, the situation is reminiscent of the strong-coupling expansion in lattice gauge theory. In that case we can also derive confinement analytically, from first principles, but not asymptotic freedom, which can only be seen at weak lattice couplings. Relaxing the $\lambda = g^2 N \gg 1$ condition on the string side would require solving the string theory beyond the supergravity approximation, and this is a difficult task, perhaps as difficult (or more difficult) than solving the gauge theory directly, without recourse to duality. One advantage that the AdS/CFT approach does have, over the lattice strong-coupling expansion, is that the universal "stringy" corrections to the static quark potential (i.e. the Lüscher and higher-order terms observed in numerical simulations) arise in a natural way, already in the supergravity approximation [21].

The AdS/CFT approach remains under very active investigation, and an important application is to the study of gauge theories in the deconfined phase. It has been argued that strongly coupled $\mathscr{N} = 4$ super Yang-Mills theory at high temperatures, which can be treated via AdS/CFT, may be a reasonable approximation to non-supersymmetric gauge theories in the deconfinement regime. An important result, derived from the gauge/string duality, is the (very low) viscosity of the quark gluon plasma [22], which seems to be in accord with experimental results [23]. This is perhaps the first time that string theory has made contact with accelerator data, and demonstrates the potential power and importance of this very intriguing approach to the physics of gauge field theories.

References

1. B. Zwiebach, *A First Course in String Theory* (Cambridge University Press, New York, 2009)
2. K. Becker, M. Becker, J.H. Schwarz, *String Theory and M-Theory: A Modern Introduction* (Cambridge University Press, New York, 2007)
3. E. Kiritsis, *String Theory in a Nutshell* (Princeton University Press, Princeton, 2007)
4. O. Aharony, S.S. Gubser, J.M. Maldacena, H. Ooguri, Y. Oz, Large N field theories, string theory and gravity. Phys. Rept. **323**, 183 (2000). [arXiv:hep-th/9905111]
5. G.T. Horowitz, J. Polchinski, Gauge/gravity duality (2006). arXiv:gr-qc/0602037
6. D. Mateos, String theory and quantum chromodynamics. Class. Quant. Grav. **24**, S713 (2007). [arXiv:0709.1523 [hep-th]]
7. R.G. Leigh, Dirac-Born-Infeld action from Dirichlet sigma model. Mod. Phys. Lett. **A4**, 2767 (1989)
8. J.M. Maldacena, The large N limit of superconformal field theories and supergravity. Adv. Theor. Math. Phys. **2**, 231 (1998). [Int. J. Theor. Phys. **38**, 1113 (1999)]. [arXiv:hep-th/9711200]
9. J.M. Maldacena, Wilson loops in large N field theories. Phys. Rev. Lett. **80**, 4859 (1998). [arXiv:hep-th/9803002]
10. E. Witten, Anti-de sitter space, thermal phase transition, and confinement in gauge theories. Adv. Theor. Math. Phys. **2**, 505 (1998). [arXiv:hep-th/9803131]
11. S.W. Hawking, D. Page, Thermodynamics of black holes in anti-DeSitter space. Comm. Math. Phys. **87**, 577 (1983)
12. A. Brandhuber, N. Itzhaki, J. Sonnenschein, S. Yankielowicz, Wilson loops, confinement, and phase transitions in large N gauge theories from supergravity. J. High Energy Phys. **9806**, 001 (1998). [arXiv:hep-th/9803263]
13. J. Polchinski, M.J. Strassler, The string dual of a confining four-dimensional gauge theory (2000). arXiv:hep-th/0003136
14. I.R. Klebanov, M.J. Strassler, Supergravity and a confining gauge theory: duality cascades and χ SB-resolution of naked singularities. J. High Energy Phys. **0008**, 052 (2000). [arXiv:hep-th/0007191]
15. J.M. Maldacena, C. Nunez, Towards the large N limit of pure $N = 1$ super Yang Mills. Phys. Rev. Lett. **86**, 588 (2001). [arXiv:hep-th/0008001]
16. P. Di Vecchia, A. Liccardo, R. Marotta, F. Pezzella, The gauge/gravity correspondence for non-supersymmetric theories. Fortsch. Phys. **53**, 450 (2005). [arXiv:hep-th/0412234]
17. T. Sakai, S. Sugimoto, Low energy hadron physics in holographic QCD. Prog. Theor. Phys. **113**, 843 (2005). [arXiv:hep-th/0412141]
18. J. Erlich, How well does AdS/QCD describe QCD? (2009). arXiv:0908.0312 [hep-ph]

19. C. Csaki, M. Reece, J. Terning, The AdS/QCD correspondence: still undelivered. J. High Energy Phys. **0905**, 067 (2009). [arXiv:0811.3001 [hep-ph]]
20. G.F. de Teramond, H.G. Dosch, S.J. Brodsky, Baryon spectrum from superconformal quantum mechanics and its light-front holographic embedding. Phys. Rev. D **91**(4), 045040 (2015). [arXiv:1411.5243 [hep-ph]]
21. O. Aharony, E. Karzbrun, On the effective action of confining strings. J. High Energy Phys. **0906**, 012 (2009). [arXiv:0903.1927 [hep-th]]
22. G. Policastro, D.T. Son, A.O. Starinets, The shear viscosity of strongly coupled $N = 4$ supersymmetric Yang-Mills plasma. Phys. Rev. Lett. **87**, 081601 (2001). [arXiv:hep-th/0104066]
23. E. Shuryak, Why does the quark gluon plasma at RHIC behave as a nearly ideal fluid? Prog. Part. Nucl. Phys. **53**, 273 (2004). [arXiv:hep-ph/0312227]

Symmetry, Confinement, and the Higgs Phase

15

Abstract

We reconsider the problem of defining confinement in theories with matter fields.

The order parameters for confinement, discussed in Chap. 4, are based on the center symmetry of the action. Confinement is the phase of unbroken center symmetry. But neither QCD, nor the SU(2) gauge Higgs theory of Eq. (3.6), are actually center symmetric. Both have matter fields in the fundamental representation of the gauge group, and the matter part of the action breaks center symmetry explicitly. Up to now we have dealt with this issue by describing QCD as a "confinement-like" theory, and the gauge Higgs theory as having confinement-like and Higgs-like regions. But this terminology is very imprecise. In the gauge Higgs theory, for example, where exactly is the boundary between the confinement-like and Higgs-like regions? And what, exactly, is the underlying physical distinction between a confinement-like and a Higgs-like theory?

QCD and the electroweak interactions are both gauge theories with matter in the fundamental representation of the gauge group. Yet the physics of these theories is clearly quite different. QCD has flux tube formation and a linear static quark potential up to string breaking; there is also a spectrum of resonances which lie on linear Regge trajectories. In the electroweak theory there is no color electric flux tube formation at any length scale, while W and Z exchange only give rise to short range Yukawa-like forces. There are two questions to be addressed. First, can the physical distinction between QCD-like theories and electroweak-like theories be formulated precisely? Secondly, in a gauge Higgs theory, is this distinction associated with broken or unbroken realization of some global symmetry? If so, then which symmetry?

The discussion in this chapter is based on recent work by K. Matsuyama and the author [1–3].

© Springer Nature Switzerland AG 2020   247
J. Greensite, *An Introduction to the Confinement Problem*, Lecture Notes
in Physics 972, https://doi.org/10.1007/978-3-030-51563-8_15

15.1   Color and Separation-of-Charge Confinement

In gauge theories with matter in the fundamental representation, large Wilson loops
have only a perimeter law falloff, and Polyakov lines are finite. A common view
is that "confinement" in such theories simply means that the asymptotic spectrum
consists only of color neutral particles. We will refer to this property as "color
confinement" or *C confinement*. As we have already discussed back in Chap. 3, this
property also holds for the asymptotic particle spectrum of gauge Higgs theories,
deep in the Higgs regime, where there is no sign of electric flux tube formation and
only Yukawa-type forces.

On the other hand, in pure SU(N) gauge theories (or in general in gauge theories
invariant under center symmetry) there is a different and stronger meaning that can
be associated with the word "confinement," which goes beyond C confinement. Of
course in such theories the particle spectrum consists of only color singlets, e.g.
glueballs in a pure gauge theory. But such theories also have the property that the
static quark potential rises linearly or, equivalently, that large planar Wilson loops
have an area-law falloff. We may ask whether there is any way to generalize this
property to gauge theories with matter in the fundamental representation.

We will begin with the fact that the Wilson area law falloff, in a pure gauge
theory, is equivalent to a property which we will call *separation-of-charge confinement*, abbreviated to S_c *confinement*. To introduce this concept, consider the process
depicted by the rectangular Wilson loop in Fig. 15.1. A static quark-antiquark pair,
connected initially by a Wilson line, evolves in Euclidean time towards a lower

Fig. 15.1 In a timelike
Wilson loop, an initial line of
color electric flux stretched
between a static quark and an
antiquark evolves, in
Euclidean time, towards a
lower energy state of the form
(15.1), represented by the red
cigar-shaped object in the
middle of the loop

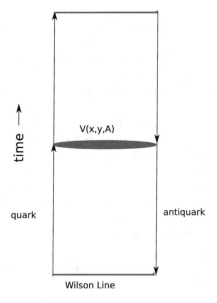

energy state of the form

$$\Psi_V \equiv \bar{q}^a(\mathbf{x})V^{ab}(\mathbf{x}, \mathbf{y}; A)q^b(\mathbf{y})\Psi_0 \,, \tag{15.1}$$

where $V(\mathbf{x}, \mathbf{y}; A)$ is a gauge bi-covariant operator transforming under a gauge transformation g as

$$V^{ab}(\mathbf{x}, \mathbf{y}; A) \rightarrow g^{ac}(\mathbf{x}, t)V^{cd}(\mathbf{x}, \mathbf{y}; A)g^{\dagger db}(\mathbf{y}, t) \,. \tag{15.2}$$

In other words, $V(\mathbf{x}, \mathbf{y}; A)$ is an operator which depends on the gauge field A, and transforms like a Wilson line between the points $\mathbf{x}, \mathbf{y}$. It is convenient to normalize V to agree with the normalization of a Wilson line, i.e.

$$\langle \Psi_0 | \mathrm{Tr}[V^\dagger(\mathbf{x}, \mathbf{y}; A)V(\mathbf{x}, \mathbf{y}; A)] | \Psi_0 \rangle = N \,, \tag{15.3}$$

where N is the number of colors. The energy above the vacuum energy $\mathscr{E}_{vac}$ is

$$E_V(R) = \langle \Psi_V | H | \Psi_V \rangle - \mathscr{E}_{vac} \,, \tag{15.4}$$

which can be obtained from a logarithmic time derivative

$$E_V(R) = -\lim_{t \to 0} \frac{d}{dt} \log \left[\frac{\langle \Psi_V | e^{-Ht} | \Psi_V \rangle}{\langle \Psi_V | \Psi_V \rangle} \right] - \mathscr{E}_{vac} \,. \tag{15.5}$$

The corresponding expression in a Euclidean time lattice gauge theory is

$$E_V(R) = -\log \left[\frac{1}{N} \langle \mathrm{Tr}[U_0^\dagger(\mathbf{x}, t)V(\mathbf{x}, \mathbf{y}, t; U)U_0(\mathbf{y}, t)V^\dagger(\mathbf{x}, \mathbf{y}, t+1, U)] \rangle \right] \,. \tag{15.6}$$

and this is what is calculated in numerical simulations. The operator $V(\mathbf{x}, \mathbf{y}, t; U)$ on the lattice depends on the spacelike link variables $U_i(\mathbf{x}, t)$ at time t.

Definition A theory has the separation-of-charge (S_c) confinement property iff

$$\lim_{R \to \infty} E_V(R) = \infty \tag{15.7}$$

for *any* choice of bi-covariant $V(\mathbf{x}, \mathbf{y}; A)$.

For an SU(N) pure gauge theory, $E_V(R) \geq E_0(R)$, where $E_0(R) \sim \sigma R$ is the ground state energy of a static quark-antiquark pair. This is the state that would be obtained after evolving the quark-antiquark pair in Fig. 15.1 for a very long period in Euclidean time.

It was proposed in [1] that S_c-confinement should also be regarded as the confinement criterion in gauge+matter theories. The crucial element is that the bi-covariant operators $V^{ab}(\mathbf{x}, \mathbf{y}; A)$ *must depend only on the gauge field A* at a fixed time, and not on the matter fields. The idea is to study the energy $E_V(R)$ in a subclass of physical states with large separations R of static color charges, unscreened by matter fields. Certainly if $V^{ab}(\mathbf{x}, \mathbf{y}; A)$ would also depend on matter field(s) in the fundamental representation of the gauge group, then it is easy to violate the S_c-confinement criterion. For example, let ϕ be a matter field in the fundamental representation, and consider

$$V^{ab}(\mathbf{x}, \mathbf{y}, \phi) = \phi^a(\mathbf{x})\phi^{\dagger b}(\mathbf{y}) , \qquad (15.8)$$

so that

$$\Psi_V = \{\overline{q}^a(\mathbf{x})\phi^a(\mathbf{x})\} \times \{\phi^{\dagger b}(\mathbf{y})q^b(\mathbf{y})\}\Psi_0 . \qquad (15.9)$$

This state corresponds to two color singlet (static quark + Higgs) states, only weakly interacting at large separations. Operators V of this kind, which depend on the matter fields, are excluded from the S_c confinement criterion. This also means that in gauge + matter theories, unlike in pure gauge theories, the lower bound $E_V(R) \geq E_0(R)$ on the energy of Ψ_V states, is *not* the lowest energy of a state containing a static quark-antiquark pair. It is the lowest energy of such states when color screening by matter is excluded.

Notice that the non-trivial center of the gauge group is essential for S_c confinement. For a gauge group such as G_2, which has a trivial center, it is always possible to construct from the gauge fields a *local* operator $\xi^a(\mathbf{x}; A)$ which transforms under a gauge transformation like a matter field in the fundamental representation. Then one replaces $\phi^a(\mathbf{x})$ by $\xi^a(\mathbf{x}; A)$ in (15.8), and again the system consists of two color singlet states, with negligible long range interaction. The fact that $\xi^a(\mathbf{x}; A)$ is a local operator is important in reaching this conclusion, which depends on the spatial separation of the operators creating the two color singlet states.

Most of the numerical investigation of S_c vs. C confinement has been done in the SU(2) gauge Higgs model of Eq. (3.6), and two questions come to mind. First, is the S_c criterion satisfied *anywhere* in the $\beta - \gamma$ phase diagram, apart from the pure gauge theory at $\gamma = 0$? The answer is yes. It can be shown that gauge-Higgs theory is S_c-confining at least in the region

$$\gamma \ll \beta \ll 1 \quad \text{and} \quad \gamma \ll \frac{1}{10} . \qquad (15.10)$$

This result is based on strong-coupling expansions and the Gershgorim Circle Theorem in linear algebra. The argument, which is lengthy, is found in [2], but intuitively the reason is that in the strong-coupling regime, subject to the limits above, string breaking takes some finite amount of Euclidean time. As a very special case, consider V chosen to be a Wilson line, and we consider an $L \times T$ Wilson loop

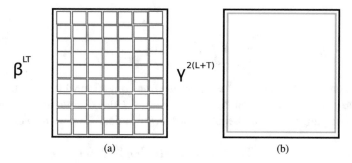

Fig. 15.2 Diagrammatic contributions to a rectangular $L \times T$ Wilson loop. (**a**) leading order in β, contributing to a linear potential; (**b**) leading order in γ, contributing to string breaking

in the SU(2) gauge Higgs theory. The leading strong coupling contributions leading to an area law falloff (plaquette term, red), perimeter law falloff (Higgs term, green) respectively, are shown in Fig. 15.2. The Wilson loop to leading order in β, γ is

$$W(L, T) = 2 \left(\frac{\beta}{4}\right)^{LT} + 2 \left(\frac{\gamma}{4}\right)^{2(L+T)} . \qquad (15.11)$$

For Euclidean times

$$T < T_{break} = 2\frac{\log \gamma}{\log \beta} , \qquad (15.12)$$

confinement dominates. Beyond this limit, the string breaks, and we have screening. Because the string-breaking time T_{break} is non-zero, the Ψ_V state with V a Wilson line will have an energy which increases linearly with separation, satisfying the S_c criterion. But of course the Wilson line state is a very special state, and S_c confinement requires that this property holds for all V. For this we refer to [2].

Given that S_c confinement holds somewhere in the $\beta - \gamma$ phase diagram, the next question is whether that property holds *everywhere* in that phase diagram, including deep in the Higgs region. The answer is no, which implies that there must be a sharp transition between the S_c and C confinement phases. Away from strong coupling there is no guarantee of S_c-confinement, and if there is even one V at some β, γ such that E_V does not grow linearly with R, then S_c-confinement is lost at that β, γ. For $V = $ Wilson line, $E_V(R) \propto R$ even for non-confining theories, but this is certainly not the only possibility. Instead we may consider, among an infinite number of possibilities, (1) the Dirac state; (2) pseudomatter states; and (3) fat link states.

The Dirac state is a generalization of the lowest energy state with static charges in an abelian theory. In an abelian theory, the gauge-invariant ground state with static $\pm$ electric charges is

$$\Psi_{\bar{q}q} = \{\bar{q}(\mathbf{x})g_C^\dagger(\mathbf{x}; A)\} \times \{g_C(\mathbf{y}; A)q(\mathbf{y})\}\Psi_0 , \qquad (15.13)$$

where

$$g_C(\mathbf{x}; A) = \exp\left[-i \int d^3z \, A_i(z)\partial_i \frac{1}{4\pi |\mathbf{x} - \mathbf{z}|}\right]. \tag{15.14}$$

It can be checked that $g_C(\mathbf{x}, A)$ is the gauge transformation which transforms A to Coulomb gauge in the abelian theory. Let $g_C(\mathbf{x}; U)$ also denote the transformation to Coulomb gauge in a non-abelian theory on the lattice. Then we may choose

$$V^{ab}(x, y; U) = g_C^{\dagger ac}(\mathbf{x}; U)g_C^{cb}(\mathbf{y}; U) \tag{15.15}$$

and

$$\Psi_V = \overline{q}^a(\mathbf{x})g_C^{\dagger ac}(\mathbf{x}; U)g_C^{cb}(\mathbf{y}; U)q^b(\mathbf{y})\Psi_0$$

$$= \overline{q}^c(\mathbf{x})q^c(\mathbf{y})\Psi_0 \quad \text{in Coulomb gauge}, \tag{15.16}$$

where the second line follows from the fact that $g_C(\mathbf{x}; U) = \mathbb{1}$ in Coulomb gauge. With the choice (15.15), the gauge invariant expression (15.6) becomes, in Coulomb gauge,

$$E_V(R) = -\log\langle\frac{1}{N}\text{Tr}[U_0(\mathbf{0}, 0)U_0^\dagger(\mathbf{R}, 0)]\rangle, \tag{15.17}$$

and this is evaluated via lattice Monte Carlo. An example of this calculation, on a sample of different lattice volumes, is shown in Fig. 15.3. Just below a critical coupling $\gamma_c \approx 0.84$, we find that $E_V(R)$ at $\beta = 2.2$ converges to a linear potential at large volume, satisfying the S_c criterion (Fig. 15.3a). Above that critical coupling, $E_V(R)$ approaches a constant asymptotically (Fig. 15.3c).

In the limit $R \to \infty$

$$\lim_{R\to\infty} E_V(R) = -\lim_{R\to\infty}\log\langle\frac{1}{N}\text{Tr}[U_0(\mathbf{0}, 0)U_0^\dagger(\mathbf{R}, 0)]\rangle$$

$$= -\log\left\{\frac{1}{N}\text{Tr}[\langle U_0(\mathbf{0}, 0)\rangle\langle U_0^\dagger(\mathbf{R}, 0)\rangle]\right\}$$

$$= \begin{cases} \text{finite if } \langle U_0\rangle \neq 0 \\ \\ \infty \quad \text{if } \langle U_0\rangle = 0 \end{cases} \tag{15.18}$$

So the transition in the Dirac state from S_c to C confining behavior corresponds to a transition in the expectation value of U_0 in Coulomb gauge, from zero to a non-zero value. This transition line was already displayed in Fig. 3.5 (upper line).

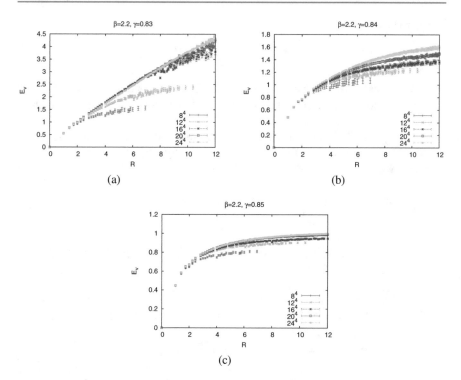

Fig. 15.3 $E_V(R)$ vs. R for the Dirac states in the gauge-Higgs model at $\beta = 2.2$. (**a**) just below the Coulomb gauge $\langle U_0 \rangle$ transition at $\gamma = 0.83$; (**b**) very close to the transition, at $\gamma = 0.84$; (**c**) just above the transition at $\gamma = 0.85$. From Greensite and Matsuyama [1]

It must be understood, however, that if $E_V(R)$ violates the condition for S_c confinement for some choice of V, then this is a sufficient condition for C confinement. But if $E_V(R)$ satisfies the S_c criterion, that is a necessary but not a sufficient condition for S_c confinement, because the condition might be violated for some other choice of V. The criterion must be satisfied for all possible choices of V, in order to be in the S_c confinement phase.

Another V operator that has been investigated is constructed from "pseudomatter" operators. These are operators dependent only the gauge field, which transform in the fundamental representation of the gauge group, i.e. like the matter fields. An example of this construction makes use of eigenstates

$$(-D_i D_i)^{ab}_{\mathbf{xy}} \varphi_n^b(\mathbf{y}) = \lambda_n \varphi_n^a(\mathbf{x}) \tag{15.19}$$

of the covariant spatial Laplacian

$$(-D_i D_i)^{ab}_{\mathbf{xy}} = \sum_{k=1}^{3} \left[2\delta^{ab}\delta_{\mathbf{xy}} - U_k^{ab}(\mathbf{x})\delta_{\mathbf{y},\mathbf{x}+\hat{k}} - U_k^{\dagger ab}(\mathbf{x} - \hat{k})\delta_{\mathbf{y},\mathbf{x}-\hat{k}} \right] . \tag{15.20}$$

We construct

$$V^{ab}(\mathbf{x}, \mathbf{y}; U) = \varphi_1^a(\mathbf{x})\varphi_1^{\dagger b}(\mathbf{y}) \tag{15.21}$$

from the lowest-lying eigenstate, and compute $E_V(R)$ by lattice Monte Carlo. This turns out to have a transition line lying close to, but a little below, that of the Dirac state.

A third possibility is based on a common noise reduction technique in lattice gauge theory known as "fat links." These are operators defined on each link by an iterative procedure

$$U_i^{(n+1)}(x) = \mathcal{N}\left\{\alpha U_i^{(n)}(x) + \sum_{j \neq i}\left(U_j^{(n)}(x)U_i^{(n)}(x+\hat{j})U_j^{\dagger}(x+\hat{i})\right.\right.$$

$$\left.\left. + U_j^{(n)\dagger}(x-\hat{j})U_i^{(n)}(x-\hat{j})U_j^{(n)}(x-\hat{j}+\hat{i})\right)\right\}, \tag{15.22}$$

starting from $U_i^1(x)$ equal to original lattice link variables, and $U_i^{fat}(x)$ is equal to the link operator obtained after the final iteration. Then we define a V operator

$$V_{fat}(x, y; A) = U_k^{fat}(x)U_k^{fat}(x+\hat{k})\dots U_k^{fat}(x+(R-1)\hat{k})$$

$$\Psi_{fat}(R) = \overline{q}(x)V_{fat}(x, y; A)q(y)\Psi_0 , \tag{15.23}$$

and compute $E_{V_{fat}}(R)$ numerically. It is found that the fat link state everywhere satisfies the S_c condition. This doesn't mean that SU(2) gauge Higgs theory is everywhere S_c-confining. It means instead that not every operator can detect the transition to a C-confinement phase.

It is difficult if not impossible to precisely identify the transition from S_c to C confinement, just by computing $E_V(R)$ for various V operators. This approach can identify lines above which the theory is definitely in the C confining phase, but one cannot be certain that the theory is in the S_c confinement region below that line. For that, a different approach is required.

15.2   The Higgs Phase as a Spin Glass

In 1975 Edwards and Anderson considered spin models with Hamiltonians of the following type [4]:

$$H_{spin} = -\sum_{xy} J_{xy}s_x s_y - h\sum_x s_x , \tag{15.24}$$

where the $s_x = \pm 1$ are Ising spins, h is a small external magnetic field, and the J_{xy} are a set of random couplings drawn from some probability distribution $P(J)$ which gives equal weight to positive and negative couplings. This is a model of a *spin glass*. It is obvious that at $h \rightarrow 0$ the model has a global Z_2 symmetry $s_x \rightarrow z s_x$, $z = \pm 1$. But because of the random nature of the couplings, the spatial average of the spins, in any thermalized configuration, will vanish on a large volume. Moreover, the expectation value of any individual spin s_x at a given spatial point will also vanish, upon averaging over all random couplings J. Nevertheless, spontaneous breaking of the global Z_2 symmetry is meaningful, and Edwards and Anderson formulated an order parameter which could detect this kind of symmetry breaking. We define

$$Z_{spin}(J) = \sum_{\{s\}} e^{-H_{spin}/kT} \tag{15.25}$$

$$\bar{s}_i(J) = \frac{1}{Z_{spin}} \sum_{\{s\}} s_i e^{-H_{spin}/kT} \tag{15.26}$$

$$q(J) = \frac{1}{V} \sum_i \bar{s}_i^2(J) \tag{15.27}$$

$$\langle q \rangle = \int \prod_{ij} dJ_{ij}\, q(J) P(J) \,. \tag{15.28}$$

$q(J)$ is the Edwards-Anderson order parameter. It is invariant with respect to Z_2 transformations $s_i \rightarrow z s_i$. Nevertheless, when

$$\lim_{h \rightarrow 0} \lim_{V \rightarrow \infty} \langle q \rangle > 0 \tag{15.29}$$

in the appropriate vanishing external field and infinite volume limits, the global Z_2 symmetry of the theory has been spontaneously broken. This is the spin glass phase. Physically it means that for a set of random couplings drawn from the $P(J)$ probability distribution, the spin configuration tends to fluctuate around some fixed configuration in which, in the appropriate limits, the average spin $\bar{s}_i(J)$ is non-zero at any site i, even though the spatial average vanishes. In this sense, the global Z_2 symmetry is broken, and that is what is detected by the Edwards-Anderson $q(J)$ order parameter.

   Like $\langle s_i \rangle$ in a spin glass model, the expectation value of a Higgs field $\langle \phi \rangle$ vanishes in a gauge Higgs theory, and for similar reasons. Obviously $\langle \phi \rangle$ must vanish, at least in the absence of gauge fixing, because ϕ is gauge covariant. But the spin glass analogy suggests another point of view. In making the spin glass analogy in the gauge Higgs context, the Higgs field plays the role of the spins, and the lattice link variables serve as the random nearest-neighbor couplings. We begin with

$$\exp[H(\phi, U)/kT] = \langle \phi, U | e^{-H/kT} | \phi, U \rangle$$
$$= \sum_n |\Psi_n(\phi, U)|^2 e^{-E_n/kT} , \tag{15.30}$$

where the Ψ_n are energy eigenstates. $H(\phi, U)$ has a vast number of near-degenerate minima at any fixed U. In analogy to spin models, we insert a small custodial symmetry breaking term

$$H_{spin}(\phi, U, \eta) = H(\phi, U) - h \sum_{\mathbf{x}} \text{Tr}[\eta^{\dagger}(\mathbf{x})\phi(\mathbf{x})] \tag{15.31}$$

with $\eta(\mathbf{x})$ an SU(2)-valued field. Then we define

$$Z_{spin}(U, \eta) = \int D\phi(\mathbf{x}) \, e^{-H_{spin}(\phi, U, \eta)/kT} \tag{15.32}$$

$$\overline{\phi}(\mathbf{x}; U, \eta) = \frac{1}{Z_{spin}(U, \eta)} \int D\phi \, \phi(\mathbf{x}) e^{-H_{spin}(\phi, U, \eta)/kT} \tag{15.33}$$

$$\Phi(U) = \frac{1}{V} \left[\sum_{\mathbf{x}} |\overline{\phi}(\mathbf{x}; U, \eta)| \right]_{\eta \in \mathcal{N}(U)} \tag{15.34}$$

$$\langle \Phi \rangle = \int DU_i(\mathbf{x}) \, \Phi(U) P(U) , \tag{15.35}$$

which should be compared to Eqs. (15.25)–(15.28). $P(U)$ is a gauge invariant probability distribution for the link variables, described below, which is obtained from the partition function after integrating out the scalar field. The expression $\mathcal{N}(U)$ represents a set $\eta(\mathbf{x})$ fields which maximize

$$\sum_{\mathbf{x}} \left| \int D\phi \, \phi(\mathbf{x}) e^{-H_{spin}(\phi, U, \eta)/kT} \right| . \tag{15.36}$$

The elements of this set at a given U transform into one another by custodial transformations. It can be shown that $Z_{spin}(U) \equiv Z_{spin}(U, \eta \in \mathcal{N}(U))$ and the order parameter $\Phi(U)$ are both gauge invariant, even at finite h, and independent of the choice of $\eta \in \mathcal{N}(U)$. Here $|\overline{\phi}(x)|$ denotes the gauge invariant modulus, e.g.

$$|\overline{\phi}(x)| = \sqrt{\text{Tr}\overline{\phi}^{\dagger}(x)\overline{\phi}(x)} \quad \text{SU(2) gauge-Higgs theory}$$

$$|\overline{\phi}(x)| = \sqrt{\overline{\phi}^{\dagger}(x) \cdot \overline{\phi}(x)} \quad \text{SU(N) gauge Higgs theory} , \tag{15.37}$$

and $P(U)$ is a gauge-invariant probability distribution for the link variables.

In a spin glass, $P(J)$ can be taken as the product of independent probability distributions for each J_{ij}, which are typically taken to be either Gaussian distributions $\exp(-\alpha J_{ij}^2)$, or else $J_{ij} = \pm J$ with equal probability for each sign, and the pairs of sites i, j are sometimes restricted to be nearest neighbors. In gauge Higgs theory, however, $P(U)$ is determined from the condition that for any gauge invariant observable $Q(U)$ defined on a time slice, we have by definition

$$
\begin{aligned}
\langle Q \rangle &= \frac{\operatorname{Tr} Q e^{-H_{spin}/kT}}{\operatorname{Tr} e^{-H_{spin}/kT}} \\
&= \frac{1}{Z} \int DU_i(\mathbf{x}) Q(U) \int D\phi(\mathbf{x}) e^{-H_{spin}(\phi,U,\eta \in \mathcal{N}(U))/kT} \\
&= \frac{1}{Z} \int DU_i(\mathbf{x})\, Q(U) Z_{spin}(U) \\
&= \int DU_i(\mathbf{x})\, Q(U) P(U) ,
\end{aligned}
\tag{15.38}
$$

and therefore

$$
P(U) = \frac{Z_{spin}(U)}{Z} ,
\tag{15.39}
$$

where

$$
Z = \int DU_i(\mathbf{x}) D\phi(\mathbf{x})\, e^{-H_{spin}(U,\phi)/kT} .
\tag{15.40}
$$

We now have a completely gauge-invariant criterion for a gauge Higgs theory to be in the spin glass phase

$$
\lim_{h \to 0} \lim_{V \to 0} \langle \Phi \rangle \begin{cases} = 0 & \text{no spin glass} \\ > 0 & \text{spin glass} \end{cases} ,
\tag{15.41}
$$

which is entirely analogous to the Edwards-Anderson spin glass criterion

$$
\lim_{h \to 0} \lim_{V \to 0} \langle q \rangle \begin{cases} = 0 & \text{no spin glass} \\ > 0 & \text{spin glass} \end{cases} .
\tag{15.42}
$$

As already mentioned, there is a global Z_2 symmetry which is spontaneously broken in the spin glass phase of the Edwards-Anderson model. The corresponding symmetry that is broken in the spin glass phase of a gauge Higgs theory is known as *custodial symmetry*.

Custodial symmetry is a symmetry of the action under a global group of transformations of the Higgs field (or Higgs fields) which does not transform the gauge field. We see, for example, that in the SU(2) gauge Higgs theory

$$
\begin{aligned}
S &= S_W[U] + S_H[\phi, U] \\
&= -\beta \sum_{plaq} \tfrac{1}{2}\mathrm{Tr}[U_\mu(x)U_\nu(x+\hat{\mu})U_\mu^\dagger(x+\hat{v})U_\nu^\dagger(x)] \\
&\quad -\gamma \sum_{x,\mu} \tfrac{1}{2}\mathrm{Tr}[\phi^\dagger(x)U_\mu(x)\phi(x+\widehat{\mu})] \,,
\end{aligned}
\tag{15.43}
$$

where ϕ is an SU(2) group-valued field, the action is invariant under the following transformations:

$$
\begin{aligned}
U_\mu(x) &\to L(x)U_\mu(x)L^\dagger(x+\hat{\mu}) \\
\phi(x) &\to L(x)\phi(x)R \,,
\end{aligned}
\tag{15.44}
$$

where $L(x) \in SU(2)_{gauge}$ is a local gauge transformation, while $R \in SU(2)_{global}$ is a global transformation. $SU(2)_{global}$ is known as the group of custodial symmetry [5]. For a U(1) or SU(3) gauge group, the custodial group would be global U(1). When the link variables $U_i(\mathbf{x})$ are fixed, then the only remaining symmetry in the system described by Z_{spin} is custodial symmetry, and this symmetry, depending on the fixed link variables, can be spontaneously broken. If the symmetry is broken for typical configurations drawn from the $P(U)$ probability distribution, then $\langle \Phi \rangle > 0$ in the appropriate limits, and custodial symmetry is spontaneously broken in the spin glass phase.

15.2.1 Numerical Evaluation

The Edwards-Anderson Hamiltonian can be easily evaluated for any set of spins $\{s_i\}$ and couplings $\{J_{ij}\}$. This is not true for $H(\phi, U)$ in (15.30). However, by the usual arguments, $H(\phi, U)$ can be expressed in terms of a path integral in Euclidean time. Identifying the arguments of H, i.e. $\phi(\mathbf{x})$, $U_i(\mathbf{x})$, as the Euclidean time dependent fields $\phi(\mathbf{x}, 0)$, $U_i(\mathbf{x}.0)$ on the $t = 0$ time slice, we have

$$
\exp[H(\phi(\mathbf{x}), U_i(\mathbf{x}))/kT] = \int DU_0 [DU_i \, D\phi]_{t\neq0} \exp[-S_E(\phi(\mathbf{x}, t), U_\mu(\mathbf{x}, t))] \,,
\tag{15.45}
$$

where S_E is the Euclidean action. The notation $[DU_i \, D\phi]]_{t\neq0}$ means that only fields at times $t \neq 0$ are integrated over.

In numerical simulations the custodial symmetry breaking term may be dispensed with, and the expectation value $\langle \Phi \rangle$ can then evaluated numerically by a Monte-Carlo-within-a-Monte-Carlo procedure. The gauge and Higgs fields are updated in the normal way for some number of iterations. The resulting gauge field on any time slice, $t = 0$ in particular, is easily seen to have been drawn from the probability distribution (15.39), and the ϕ field can be taken as a starting configuration in the data taking process. Data taking involves a series of n_{sym} sweeps through the $D = 4$ dimensional lattice, updating the ϕ field and the link variables at times $t \neq 0$ in the usual way. The link variables at $t = 0$ are held fixed. Let $\phi(\mathbf{x}, t = 0, n)$ be the scalar field at site $\mathbf{x}$ on the $t = 0$ timeslice at the n-th sweep. Then we compute

$$\Phi(n_{sym}) = \frac{1}{V} \sum_{\mathbf{x}} |\overline{\phi}(\mathbf{x})| , \qquad (15.46)$$

where

$$\overline{\phi}(\mathbf{x}) = \frac{1}{n_{sym}} \sum_{n=1}^{n_{sym}} \phi(\mathbf{x}, 0, n) , \qquad (15.47)$$

and averaging over the Φ's obtained at each data taking interval gives an estimate for $\langle \Phi(n_{sym}) \rangle$. $\langle \Phi(n_{sym}) \rangle$ is then extrapolated to $n_{sym} \to \infty$ by fitting to an expression which is expected on general statistical grounds

$$\langle \Phi(n_{sym}) \rangle = \langle \Phi_\infty \rangle + \frac{\text{constant}}{\sqrt{n_{sym}}} . \qquad (15.48)$$

Transition points are determined from the points at fixed β, γ where $\langle \Phi_\infty \rangle$ begins to move away from zero as γ increases. Examples of the extrapolation to $n_{sym} \to \infty$ is shown at $\beta = 1.2$ above ($\gamma = 1.5$) and below ($\gamma = 1.25$) the transition are shown in Fig. 15.4a. The custodial symmetry breaking/spin glass transition line is displayed in Fig. 15.4b, and compared with remnant symmetry breaking transition line in Coulomb gauge.

15.3   The Spin Glass and the S_c-to-C Transitions

The spin glass phase is the phase of spontaneously broken custodial symmetry. The transition to this phase coincides, in the absence of a massless phase, with the transition from S_c to C confinement. To justify this statement, we consider quantization of gauge Higgs theories with gauge conditions $F(U) = 0$ involving only spacelike links, which can be imposed independently on each time slice, and which leave unfixed only a global subgroup of the gauge group on any given timeslice. We will refer to these as *F-gauges*, examples are Coulomb and maximally fixed axial gauges, and they are special in the sense that field operators $q, \overline{q}, \phi$ transformed to an F-gauge, operating on the ground state, generate physical states.

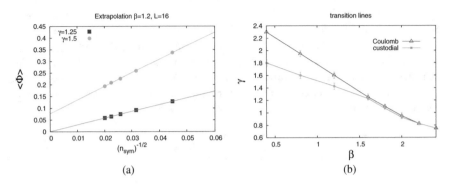

Fig. 15.4 (a) Extrapolation of $\langle\Phi\rangle$ to $n_{sym} \rightarrow \infty$ above ($\gamma = 1.5$) and below ($\gamma = 1.25$) the custodial symmetry breaking transition at $\beta = 1.2, \gamma = 1.4$. Error bars are smaller than the symbol sizes. (b) Transition lines separating the unbroken (smaller γ) and broken (larger γ) phases of (1) remnant gauge symmetry in Coulomb gauge (triangular data points), and (2) custodial symmetry (filled circles). Note that the broken Coulomb phase lies entirely within the broken custodial symmetry phase, as it must from the bound in (15.50). From Greensite and Matsuyama [3]

Any F-gauge leaves some global gauge symmetry unfixed. This remnant gauge symmetry includes at a minimum, for the SU(N) gauge group,

$$\phi(\mathbf{x}, t) \rightarrow z(t)\phi(\mathbf{x}, t) \quad , \quad U_0(\mathbf{x}, t) \rightarrow z(t)U_0(\mathbf{x}, t)z^\dagger(t+1) \quad , \quad z(t) \in Z_N ,$$
$$(15.49)$$

and because this is a global symmetry it can be spontaneously broken on a time slice. Some gauges, e.g. Coulomb gauge, leave unfixed a global SU(N) symmetry, others may leave only a subgroup composed of diagonal matrices, and there exist some gauges which leave only the global remnant symmetry (15.49).

Let $|\langle\phi\rangle_F|$ be the modulus of the scalar field expectation value evaluated in an F-gauge. It can be shown that

1. for all physical F-gauges, $\langle\Phi\rangle \geq |\langle\phi\rangle|_F$ (15.50)

2. there exists at least one F gauge such that $\langle\Phi\rangle = \langle\phi\rangle_F$.   (15.51)

This means that broken custodial symmetry is a necessary condition for $\langle\phi\rangle_F \neq 0$ in any F-gauge, and a sufficient condition for $\langle\phi\rangle_F \neq 0$ in at least one F gauge. The proofs of these statements are found in ref. [3].

If custodial symmetry is unbroken, then it implies as a special case that

$$\langle\Psi_0|Q(z\phi, U)|\Psi_0\rangle = \langle\Psi_0|Q(\phi, U)|\Psi_0\rangle \quad , \quad z(t) \in Z_N .$$
$$(15.52)$$

But if this holds true for the center subgroup of custodial symmetry, it must also hold true for the Z_N subgroup of remnant global gauge symmetry, since these transformations have exactly the same effect on the Higgs field and spacelike links on a time slice. Likewise, broken custodial symmetry implies broken remnant gauge invariance, and this is true in all F-gauges. This correspondence between the center subgroups of custodial and remnant gauge symmetry has important consequences; in particular it means that the remnant gauge symmetry is broken in the spin glass phase.

Remnant gauge symmetry is useful in identifying states which contain isolated charges. A good example is provided by the abelian theory, where the operator

$$\phi_C(\mathbf{x}) = g_C(\mathbf{x}; A)\phi(\mathbf{x}) \,, \tag{15.53}$$

applied to the vacuum, creates a state of unit electric charge, where g_C is the transformation to Coulomb gauge explicitly in (15.14). Under an arbitrary gauge transformation $g(\mathbf{x}) = e^{i\theta(\mathbf{x})}$, the Coulomb gauge operator $\phi_C(\mathbf{x})$ transforms as $\phi_C(\mathbf{x}) \rightarrow e^{i\theta_0}\phi_C(\mathbf{x})$, where θ_0 is the $k = 0$ mode of the Fourier transformed $\theta(k)$. In other words, ϕ_C transforms under the remnant global symmetry in Coulomb gauge. This operator, when applied to the vacuum, creates a state with the same transformation property, providing the vacuum is itself invariant under this remnant gauge symmetry. This idea extends to non-abelian lattice gauge theory. Let $g_F(\mathbf{x}; U)$ be the transformation to an F, and we define

$$\phi_F(\mathbf{x}) = g_F(\mathbf{x}, U)\phi(\mathbf{x})$$
$$q_F(\mathbf{x}) = g_F(\mathbf{x}, U)q(\mathbf{x})$$
$$\overline{q}_F(\mathbf{x}) = \overline{q}(\mathbf{x})g_F^\dagger(\mathbf{x}, U)$$
$$U_{i,F}(\mathbf{x}) = g_F(\mathbf{x}, U)U_i(\mathbf{x})g_F^\dagger(\mathbf{x}+\hat{\imath}, U) \,. \tag{15.54}$$

The $q_F, \overline{q}_F, \phi_F$ field operators are invariant under local gauge transformations, but still transform covariantly under global transformations in the remnant gauge group (15.49), and therefore create charged states when operating on the vacuum, providing the ground state is invariant under the remnant symmetry.

However, the ground state in the spin glass phase is not invariant under that global symmetry. In this phase there is always at least one F-gauge in which $\langle\phi\rangle_F \neq 0$, which implies that the remnant gauge symmetry (15.49) is spontaneously broken, and the ground state is not a state of definite (zero) color charge, in the sense that it is neither invariant nor covariant under the remnant gauge symmetry. In this phase, the color electric field created by the field operators is of finite energy. To see this, we first note that $\langle\phi\rangle_F \neq 0$ implies that also $\langle U_0\rangle_F \neq 0$. The reason is that $U_0(\mathbf{x}, t)$ transforms under a global $Z_N \times Z_N$ symmetry at times $t, t+1$. If ϕ fluctuates around a fixed background on every time slice, then this background breaks the $Z_N \times Z_N$, and U_0 will acquire a non-vanishing expectation value. This is sufficient to show that the spin glass phase is not an S_c confining phase, which is seen by choosing

$$V_F(\mathbf{x}, \mathbf{y}, t; U) = g_F^\dagger(\mathbf{x}, t; U)g_F(\mathbf{y}, t; U) \,, \tag{15.55}$$

so that

$$\Psi_V(R) = \overline{q}_F^a(\mathbf{x})q_F^a(\mathbf{y})\Psi_0$$
$$= \overline{q}^a(\mathbf{x})V_F^{ab}(\mathbf{x}, \mathbf{y}, t; U)q^b(\mathbf{y})\Psi_0 . \tag{15.56}$$

Computing $E_V(R)$ in the F-gauge, where $V_F = \mathbb{1}$ and $\langle U_0 \rangle \neq 0$, we find as before

$$\lim_{R \to \infty} E_V(R) = -\lim_{R \to \infty} \log\left[\frac{1}{N}\langle \text{Tr}[U_0^\dagger(\mathbf{x}, t)U_0(\mathbf{y}, t)]\rangle_F\right]$$

$$= -\log\left[\frac{1}{N}\text{Tr}[\langle U_0^\dagger(\mathbf{x}, t)\rangle_F \langle U_0(\mathbf{y}, t)\rangle_F]\right]$$

$$= \text{finite} , \tag{15.57}$$

which violates the S_c confinement condition.

Moreover, because the vacuum is not a state of definite (zero) color charge, there is no essential distinction between charged and neutral states

$$|\text{charged}_{\mathbf{xy}}\rangle = \overline{q}^a(\mathbf{x})V_F^{ab}(\mathbf{x}, \mathbf{y}; U)q^b(\mathbf{y})|\Psi_0\rangle$$

$$|\text{neutral}_{\mathbf{xy}}\rangle = (\overline{q}^a(\mathbf{x})\phi^a(\mathbf{x}))(\phi^{\dagger a}(\mathbf{y})q^a(\mathbf{y}))|\Psi_0\rangle , \tag{15.58}$$

where we imagine taking $\mathbf{y} \to \infty$, leaving an isolated quark at site $\mathbf{x}$. These are both physical states, but the neutral state is obtained from operators creating two separated color singlet objects, with no color electric field diverging from points $\mathbf{x}, \mathbf{y}$. The charged state is created by operators which transform, at sites $\mathbf{x}, \mathbf{y}$, under the remnant gauge symmetry. Then evaluating the overlap in the F-gauge, where $V_F = \mathbb{1}$, integrating out the heavy quark fields, and taking the $R = |\mathbf{x} - \mathbf{y}| \to \infty$ limit

$$\lim_{|\mathbf{x}-\mathbf{y}| \to \infty} \langle \text{neutral}|\text{charged}\rangle \propto \lim_{|\mathbf{x}-\mathbf{y}| \to \infty} \langle \phi^{\dagger a}(\mathbf{x})\phi^a(\mathbf{y})\rangle_F$$

$$= \langle \phi^{\dagger a}\rangle_F \langle \phi^a\rangle_F$$

$$> 0 . \tag{15.59}$$

Since there is a non-zero overlap between the charged state containing an isolated quark at site $\mathbf{x}$, and the neutral state created by a color singlet quark-scalar operator at site $\mathbf{x}$, we conclude that the "charged" state is not really charged; it has a finite overlap with states created by color singlet operators acting on the vacuum at point $\mathbf{x}$. If the state created by a color singlet operator is neutral, then so is the state created by the charged operator. In fact, we see that the "charged" state is not associated with a long range color electric field characteristic of a charged field. If it were, and because there is no such long range field in the state created by a color singlet operator acting on the vacuum, then the charged and neutral states would be orthogonal, which is not the case. Because there are no color charged states in the spectrum which can be distinguished, by their symmetry properties (or by a long range color electric field) from neutral states, we conclude that the spin glass phase is a phase of C confinement.

On the other hand, there are also F gauges in which $\langle \phi \rangle_F = 0$ in at least some part of the spin glass phase. It makes no sense to say that the spin glass phase is C confining in some gauges but not in others, and in fact the spontaneous breaking of custodial symmetry, which implies spontaneous breaking of remnant gauge symmetry, is independent of any gauge choice, either explicit or implicit. The resolution of this conundrum must be that orthogonality of the state $\Psi = \phi_F(\mathbf{x})\Psi_0$ with the vacuum state, i.e. $\langle \phi \rangle_F = \langle \Psi_0 | \Psi \rangle = 0$ is not a consequence of unbroken remnant gauge symmetry, but is rather due to the fact that the energy of Ψ, above the vacuum energy, is infinite. Following the line of argument in (15.57), we see that a field operator creates a state of infinite energy if $\langle U_0 \rangle_F = 0$ in such F gauges. A good example, where this resolution can be seen explicitly, is axial gauge. In that gauge it can be shown that $\langle \phi \rangle_F = \langle U_0 \rangle_F = 0$ everywhere in the phase diagram, including in the spin glass region. The underlying reason is that in this gauge the field operator creates a line of electric flux stretching to infinity, and this is a state of infinite energy in any phase.

Unbroken custodial symmetry implies unbroken remnant gauge symmetry, and states can be classified according to their transformation properties under the global center subgroup of the gauge symmetry. Charged and neutral states, in this phase, are orthogonal. In particular, let us return to the overlap of charged and neutral states, as defined in (15.58), this time in the symmetric phase, for any choice of $V(\mathbf{x}, \mathbf{y}; U)$ operator. We have

$$\lim_{|\mathbf{x}-\mathbf{y}|\to\infty} \langle \text{neutral}|\text{charged} \rangle \propto \lim_{|\mathbf{x}-\mathbf{y}|\to\infty} \langle \phi^{\dagger a}(\mathbf{x}) V^{ab}(\mathbf{x}, \mathbf{y}; U) \phi^b(\mathbf{y}) \rangle$$

$$= \lim_{|\mathbf{x}-\mathbf{y}|\to\infty} \int DU \overline{\phi^a(\mathbf{x})\phi^b(\mathbf{y})}[U] V^{ab}(\mathbf{x}, \mathbf{y}; U) P(U) ,$$

$$(15.60)$$

where

$$\overline{\phi^a(\mathbf{x})\phi^b(\mathbf{y})}[U] = \frac{1}{Z_{spin}(U)} \int d\phi \phi^a(\mathbf{x})\phi^b(\mathbf{y}) e^{-H_{spin}(U,\phi)/kT} . \quad (15.61)$$

Since custodial symmetry is unbroken for gauge configurations drawn from the probability distribution $P(U)$, it follows that for such configurations, in the symmetric phase at $h \to 0$,

$$\lim_{|\mathbf{x}-\mathbf{y}|\to\infty} \overline{\phi^a(\mathbf{x})\phi^b(\mathbf{y})}[U] = 0 . \quad (15.62)$$

Because V is a bounded operator (see (15.3)), this means that the overlap between all charged states and the neutral, "string broken" states must vanish in the $R \to \infty$ limit:

$$\lim_{|\mathbf{x}-\mathbf{y}|\to\infty} \langle \text{neutral}|\text{charged}\rangle = 0 . \tag{15.63}$$

and it should be emphasized that this result holds for all V operators in the symmetric phase, independent of gauge choice.

Charged states in the symmetric phase may be of either finite or infinite energy above the ground state energy. As an example, it could be that in some F gauge in the symmetric phase we have $\langle U_0 \rangle_F \neq 0$, where it should be noted that although $\langle \phi \rangle_F \neq 0$ implies $\langle U_0 \rangle_F \neq 0$, the converse is not necessarily true. In this case, by previous arguments, the field operators create charged states of finite energy. If there are finite energy charged states, orthogonal to all neutral states, then states of that kind must appear in the asymptotic particle spectrum, and the system cannot be C confining. Also the system cannot be in an S_c confining phase, where isolated charges are of infinite energy. The remaining possibility is a massless phase. Such phases exist e.g. in the abelian Higgs model with a compact gauge group in 3+1 dimensions [6], and in SU(2) lattice gauge theory in 4+1 dimensions [7].

So the custodial symmetric phase is either S_c confining or massless. This is consistent with the fact that $\langle \phi \rangle$ in the symmetric phase vanishes in all physical F-gauges, and there is no sensible perturbative expansion around a non-zero value of ϕ in this class of gauges. There is no broken gauge symmetry of any kind in this phase. The conclusion is that the spin glass phase is a C confinement Higgs phase, while the phase of unbroken custodial symmetry may be either a massless or an Sc confining phase, depending on the couplings and on the spacetime dimension. In the absence of a massless phase, as in SU(2) gauge Higgs theory in D = 3 + 1 dimensions, the transition to the spin glass phase and the transition from S_c to C confinements coincide.

To summarize, the S_c condition and the spin glass order parameter provide us with a more precise understanding of what the word "confinement" means in theories with matter in the fundamental representation of the gauge group, where the order parameters discussed in Chap. 4 do not apply. Confinement, or more precisely separation-of-charge confinement, means that (1) the system is in a phase with an unbroken global center symmetry which distinguishes between color charged and color neutral states; but that (2) the energy of color charged states above the vacuum energy is infinite, so that isolated color charged particles never appear in the spectrum. The Higgs phase is a phase of broken custodial symmetry, and can be understood as a C confinement spin glass phase. Separation-of-charge confinement, as opposed to color confinement, is only meaningful for gauge groups which contain a non-trivial center subgroup.

References

1. J. Greensite, K. Matsuyama, Confinement criterion for gauge theories with matter fields. Phys. Rev. D **96**(9), 094510 (2017). [arXiv:1708.08979 [hep-lat]]
2. J. Greensite, K. Matsuyama, What symmetry is actually broken in the Higgs phase of a gauge-Higgs theory? Phys. Rev. D **98**(7), 074504 (2018). [arXiv:1805.00985 [hep-th]]
3. J. Greensite, K. Matsuyama, Higgs phase as a spin glass and the transition between varieties of confinement. Phys. Rev. D **101**, 054508 (2020). [arXiv:2001:03068 [hep-th]]
4. S.F. Edwards, P.W. Anderson, Theory of spin glasses. J. Phys. F Metal Phys. **5**, 965 (1975)
5. A. Maas, Brout-Englert-Higgs physics: from foundations to phenomenology. Prog. Part. Nucl. Phys. **106**, 132 (2019). [arXiv:1712.04721 [hep-ph]]
6. J. Ranft, J. Kripfganz, G. Ranft, Phase structure, magnetic monopoles and vortices in the lattice Abelian Higgs model. Phys. Rev. D **28**, 360 (1983)
7. B.B. Beard, R.C. Brower, S. Chandrasekharan, D. Chen, A. Tsapalis, U.-J. Wiese, D-theory: field theory via dimensional reduction of discrete variables. Nucl. Phys. Proc. Suppl. **63**, 775 (1998). [hep-lat/9709120]

Concluding Remarks

The ideas surveyed in previous chapters do not exhaust all of the suggestions that have been made over the years, regarding the confinement mechanism. To include every proposal would have required an encyclopedia, not an introduction to the subject. I have included those ideas which seem to me to be the most promising, and the most likely to survive as part of our eventual understanding of confinement. But the selection comes without any guarantees. The solution of the confinement problem may eventually emerge from a further development of ideas which have been presented here. It may also come from some other direction entirely.

I have tried to stress, in this volume, a few main themes. Among these are the importance of carefully defining the problem that we are actually trying to solve, and the relevance of the center of the gauge group. The word "confinement" means different things to different people, and the existence of a color singlet spectrum and a mass gap is a feature not just of QCD, but also of Higgs theories having purely Yukawa-like forces. A color-singlet spectrum may have little to do with linear potentials and color electric flux tube formation, and if we are interested in the latter properties, it is better to focus on the asymptotic behavior of Wilson loops in the vacuum state. A vacuum state in which Wilson loops have an area-law falloff for arbitrarily large loops is qualitatively different, in its large-scale structure, from a vacuum state which does not have this property, and the symmetry which distinguishes between the two different types is an unbroken, and non-trivial, global center symmetry.

A strong clue about the nature of magnetic disorder responsible for the confining force is the dependence of the string tension on the color representation of the static charges. Asymptotically, the string tension depends only on the N-ality of the group representation, suggesting that vacuum fluctuations must arrange themselves, at large scales, in such a way that a Wilson loop is affected only via its N-ality. The existence of center vortices, or at least of domains whose color magnetic flux is quantized in units of the center subgroup, seems for this reason almost inevitable. At intermediate distance scales, extending to infinity as the number of colors $N \to \infty$,

© Springer Nature Switzerland AG 2020

J. Greensite, *An Introduction to the Confinement Problem*, Lecture Notes in Physics 972, https://doi.org/10.1007/978-3-030-51563-8_16

the string tension appears instead to be proportional to the quadratic Casimir of the color charge representation. A theory of confinement, in terms of the field configurations which dominate the vacuum state, must ultimately account for not only the asymptotic dependence on N-ality, but also the Casimir scaling found at intermediate distances, as well as the transition between the two regimes.

Not every proposed explanation of confinement is based on the nature of vacuum field configurations (such as vortices, monopoles, and calorons). The Green's function approach, which relies on solutions of the Dyson-Schwinger equations, the Gribov-Zwanziger scenario, and especially the AdS/CFT approach, relating gauge theory to a dual theory of strings in ten-dimensional space-time, are important counter-examples. It is, of course, possible that confinement can be understood in more than one way, and it could happen that more than one approach will ultimately succeed. We would then expect the successful explanations to be related to one another, and somehow interdependent.

Confinement, in any case, is the hard problem of hadronic physics. It is also the hard problem for non-abelian gauge theories in general. Asymptotic freedom at short distances has been understood since the 1970s, but a similar grasp of the confining property of non-abelian gauge theories at large distances has so far eluded us. A number of ideas about confinement, introduced in previous chapters, seem well-motivated and promising in various degrees. But until non-abelian gauge theories are solved analytically, there is likely to be disagreement about the structure of the vacuum state, the origin of the mass gap, and the origin of the confining force. While there has been a great deal of progress, as outlined in this volume, the confinement problem is still open. As long as this problem remains open, a large gap will also remain in our understanding of the strongest of the known fundamental forces in Nature.

Index

© Springer Nature Switzerland AG 2020
J. Greensite, *An Introduction to the Confinement Problem*, Lecture Notes
in Physics 972, https://doi.org/10.1007/978-3-030-51563-8